AF846495

# Code Biology

Marcello Barbieri

# Code Biology

## A New Science of Life

Springer

Marcello Barbieri
University of Ferrara
Dipartimento di Morfologia ed Embriologia
Ferrara, Italy

ISBN 978-3-319-14534-1 ISBN 978-3-319-14535-8 (eBook)
DOI 10.1007/978-3-319-14535-8

Library of Congress Control Number: 2014959281

Springer Cham Heidelberg New York Dordrecht London
© Springer International Publishing Switzerland 2015
This work is subject to copyright. All rights are reserved by the Publisher, whether the whole or part of the material is concerned, specifically the rights of translation, reprinting, reuse of illustrations, recitation, broadcasting, reproduction on microfilms or in any other physical way, and transmission or information storage and retrieval, electronic adaptation, computer software, or by similar or dissimilar methodology now known or hereafter developed.
The use of general descriptive names, registered names, trademarks, service marks, etc. in this publication does not imply, even in the absence of a specific statement, that such names are exempt from the relevant protective laws and regulations and therefore free for general use.
The publisher, the authors and the editors are safe to assume that the advice and information in this book are believed to be true and accurate at the date of publication. Neither the publisher nor the authors or the editors give a warranty, express or implied, with respect to the material contained herein or for any errors or omissions that may have been made.

Printed on acid-free paper

Springer International Publishing AG Switzerland is part of Springer Science+Business Media (www.springer.com)

*To my wife*
*Leopoldine Margarethe*
*the girl from Klagenfurt*
*who fell in love with Italy*
*and saved my life*

# Foreword

You approach a set of red traffic lights and you stop. Why? You have read the previous sentence and understood its meaning. How? It is because you have learnt how to associate symbols with their conventional meanings—you understand a particular encoded relationship. These are examples of cultural codes, but since the 1960s we know that a true molecular code exists in every cell: the molecular code called the genetic code, which translates the four-letter nucleotide sequence of DNA into the 20-letter amino acid sequence in proteins. What Marcello Barbieri will tell you in this ground-breaking book is that between the development of the genetic code at the beginning of life on earth and the development of cultural codes and languages some four billion years later many other organic codes were established. This is in itself already a major discovery, but its implications are perhaps even more important, going right to the heart of biology. The establishment of each of these organic codes opened up a new world of possibilities for natural selection to explore, possibilities that did not exist before. We can link each of the major transitions in evolutionary history to an organic code. In short, Barbieri will give you a new lens through which to view life and its evolution. Whether in the end you accept this new code view of life will prove to be immaterial—this book will have challenged your current view of biology and may have even changed it irrevocably. As did the discovery of Barbieri's work a decade ago change mine.

I am a biochemist morphed into a systems biologist, or, as I prefer it, a molecular cell physiologist. Since the early 1980s I have been involved in a quest to understand the functional organisation and regulation of cellular processes, especially of metabolism. In 2004, I was on a year-long sabbatical, attempting to integrate the approaches of Robert Rosen, John von Neumann and Howard Pattee into a coherent picture of the cell as a self-fabricating system. I had come to this point both through my research on metabolic control and regulation and by reading Rosen's *Life Itself*, a book that changed my view of biology profoundly. But I felt that a piece of the puzzle was missing. What bothered me as a biochemist was that the link between genotype and phenotype was not made clear enough, whether in Rosen's metabolism-repair systems, or in Von Neumann's distinction between a self-reproducing system and its description, or in Pattee's matter-symbol complementarity. Of course all three of these brilliant scientists were perfectly aware of the connection, but did not make it explicit and this set me off on a search for the "missing link". I found the answer during a search for definitions of life. *The Organic Codes: An Introduction to Semantic Biology* by Marcello Barbieri contains a compendium of 63 such definitions. However, my interest in this set of definitions was eclipsed by the excitement of reading the book itself. Here was exactly what I was looking for. The key was Barbieri's concept of the ribotype as the necessary link between the two independent worlds of genotype and phenotype. As Michael Ghiselin said in his foreword to the book, "Previous scenarios treated proteins or DNA as coming first. Both of these alternatives ran into difficulties because the one cannot exist without the other. For that very reason there must have been something additional to genotype and phenotype, which he calls the ribotype. It is RNA that bridges the gap between genotype and phenotype, and it does so by endowing the system with meaning." Meaning? What was "meaning" doing in a biological text? In the biological paradigm I grew up in such a question could not even be asked, and if it was, would be laughed out of the room. Nevertheless, I was hooked, and this question and his answer to it set me off on an epistemological adventure that eventually led to my meeting Marcello, becoming a close friend and collaborator, being welcomed into the biosemiotics community, and, finally, being part in 2013 of the creation of the International Society of Code Biology and attending its first conference in 2014. I have therefore been privy to the creation of the book that you are holding in your hands, a book that will be the standard text of this new discipline for years to come. In it you will find the answer to "What is organic meaning?" and, paraphrasing Woody Allen, to many other questions you were afraid to ask.

*Code Biology: A New Science of Life* is a remarkable record of one man's quest to understand life from the viewpoint of codes. In the modern era of big science, research consortia, and the race for funding it is rare to find someone who, over solitary decades, pieces together strands of evidence and avenues of thought into a grand synthesis such as the one presented here. One can follow the thread: it started in the 1970s with Barbieri's proposal of a primitive ribosome model developed while at the Max Planck Institute for Molecular Genetics in Berlin (where Ada Yonath, who won the Nobel prize for chemistry in 2009 for her structural studies of the ribosome, worked alongside him). This model led to his ribotype theory on the origin of life published in 1981 and his 1985 book *The Semantic Theory of Evolution*. However, biology was in the thick of the recombinant DNA era and took no notice. Two more decades passed before Barbieri came to the attention of the biosemiotics community when he sent a first version of *The Organic Codes* to Thomas Sebeok, one of the fathers of modern biosemiotics. How the story unfolds from there on is told in Chap. 9. The ensuing relationship was very productive and lasted a number of years, but there was always an underlying discomfort that ultimately led to a parting of ways. Barbieri insisted from the start that organic meaning results from a mechanistic process of decoding, whereas the Peircian biosemiotic view was that all meaning is produced by interpretation. Barbieri refused to relinquish his strict adherence to a scientific approach, to relinquish the belief that the understanding of organic codes and organic meaning must be in terms of scientific models that can be tested.

I expect that a viewpoint as strong as Barbieri's is bound to be controversial in some circles. If you ascribe to the physicalist's "life is just chemistry", the molecular biologist's "life is chemistry plus information", or the Peircian biosemiotician's "the units of life are signs"—views that are described in Chap. 1—you may be rankled by the suggestion that these views are inadequate for explaining life. But they *are* inadequate, as Barbieri argues convincingly. The key lies in his realisation that material life as we know it is "artifact-making", that organisms are fabricators, ultimately of themselves. This was also Robert Rosen's main conclusion; his analysis in terms of Aristotelean causes led him to the statement that organisms are "closed to efficient causation", by which he meant that organisms autonomously manufacture all the molecular machinery necessary for life. The closure follows because these machines can produce the polynucleotides and proteins from which they themselves are formed. As explained by Barbieri, each of these two classes of highly specific biopolymers requires its own template-directed manufacturing process: copying for polynucleotides and coding for proteins. These two processes in turn form the basis for two types of evolutionary process: evolution by natural selection through copying and evolution by natural conventions through coding. We of course know and accept the mechanism of natural selection through copying as the basis for Darwinian evolution and the creation of relative novelties, as Barbieri calls them, but the coming into existence of absolute novelties has been a problem from the start. That we no longer have to resort to hand-waving arguments to explain absolute novelties is one of the main contributions of the new science of code biology: evolution by natural conventions through new organic codes provides the required mechanism.

Code biology unearths many other riches besides a novel mechanism of evolution. In the journey from life before the cell through all the major stages of evolution to the origins of human language, Barbieri touches on a wide range of subjects, among which each reader will find something to marvel over or to argue with. I found his treatment of organic information and organic meaning in terms of nominable entities a revelation. As Barbieri explains in Chap. 2, nominable entities are a new class of physical observables, objective but not measurable. The simplest example is a linear sequence, which can be described only by naming the order and identity of its constituents. Chemical molecules are also nominable entities: one has to both list the constituent atoms and describe how they are bound to each other. What is crucial is that Barbieri shows that nominable entities, and therefore organic information and organic meaning, are new *fundamental* physical observables, the discovery of which has always led to major advances in physics.

I should prepare you for something, though. Unlike many scientists who write mostly in the disembodied passive and like to hedge their bets, Barbieri writes in the classic style, a style beautifully described by Francis-Noël Thomas and Mark

Turner in their *Clear and Simple as the Truth*. In *The Sense of Style* Steven Pinker writes: "The guiding metaphor of classic style is seeing the world. The writer can see something that the reader has not yet noticed, and he orients the reader's gaze so that she can see it for herself. The purpose of writing is presentation, and its motive is disinterested truth. It succeeds when it aligns language with the truth, the proof of success being clarity and simplicity. The truth can be known, and is not the same as the language that reveals it; prose is a window onto the world. The writer knows the truth before putting it into words; he is not using the occasion of writing to sort out what he thinks. Nor does the writer of classic prose have to argue for the truth; he just needs to present it. That is because the reader is competent and can recognize the truth when she sees it, as long as she is given an unobstructed view. The writer and the reader are equals, and the process of directing the reader's gaze takes the form of a conversation." I can think of no better way of describing Barbieri's prose. Enjoy it.

Wissenschaftskolleg zu Berlin, Berlin, Germany Jannie Hofmeyr
November 2014

*on sabbatical leave from*
the Department of Biochemistry
and the Centre for Studies in Complexity
of the University of Stellenbosch, South Africa

# Introduction

The genetic code appeared on Earth at the origin of life, and the codes of culture arrived almost four billion years later. For a long time it has been assumed that these are the only codes that exist in Nature, and if that were true we would have to conclude that codes are *extraordinary exceptions* that appeared only at the beginning and at the end of the history of life. In reality, various other organic codes (codes between organic molecules) have been discovered in the past few decades.

In 1975, the American biochemist Gordon Tomkins published a paper entitled '*The Metabolic Code. Biological symbolism and the origin of intercellular communication*' (Tomkins 1975). That was the first announcement of a new organic code after the discovery of the genetic code, but tragically Tomkins died that very year and his new world of biological symbolism remained unexplored.

In 1979, David Elder pointed out that the formation of body segments in annelid worms is described by combinatorial rules that in electronics are known as *Gray code*, and proposed that the annelid body plan is based on a biological version of those rules that he referred to as *epigenetic code* (Elder 1979).

At the end of the 1980s, Edward Trifonov started a life-long campaign in favour of the idea that genomes simultaneously carry several overlapping codes, not just the genetic code, and gave them the collective name of *sequence codes* (Trifonov 1987, 1989, 1996, 1999).

Finally, at the end of the 1990s and in the early 2000s, a wide variety of new organic codes came to light. Among them: the *adhesive code* (Redies and Takeichi 1996; Shapiro and Colman 1999), the *splicing codes* (Barbieri 1998, 2003; Fu 2004; Matlin et al. 2005; Pertea et al. 2007; Wang and Burge 2008; Barash et al. 2010; Dhir et al. 2010), the *signal transduction codes* (Barbieri 1998, 2003), the *histone code* (Strahl and Allis 2000; Jenuwein and Allis 2001; Turner 2000, 2002, 2007; Kühn and Hofmeyr 2014), the *sugar code* (Gabius 2000, 2009), the *compartment codes* (Barbieri 2003), the *cytoskeleton codes* (Barbieri 2003; Gimona 2008), the *tubulin code* (Verhey and Gaertig 2007), the *nuclear signalling code* (Maraldi 2008), the *apoptosis code* (Basañez and Hardwick 2008; Füllgrabe et al. 2010), the *ubiquitin code* (Komander and Rape 2012), the *bioelectric code* (Tseng and Levin 2013; Levin 2014), the *glycomic code* (Buckeridge and De Souza 2014) and the *acoustic codes* (Farina and Pieretti 2014).

It must be underlined that codes have been defined in different ways, a problem that is not uncommon in biology, but in our case this is not an insurmountable obstacle because there is an operative definition that can be applied to all organic codes. This definition, furthermore, has been instrumental in the development of mathematical models that can estimate the presence of organic codes in natural systems (De Beule et al. 2011; Görlich et al. 2011; Görlich and Dittrich 2013).

## An Operative Definition

An operative definition is one that allows us to make experimental tests which prove whether or not organic codes exist in Nature. The starting point is the idea that a code is always *a set of rules that establish a correspondence* (or a *mapping*) *between two independent worlds* (Barbieri 2003). The Morse code, for example, is a mapping between the letters of the alphabet and groups of dots and dashes. The highway code is a correspondence between street signals and driving behaviours (a red light means 'stop', a green light means 'go', and so on).

What is essential in all codes is that the coding rules, although completely compatible with the laws of physics and chemistry, are not dictated by these laws. In this sense they are *arbitrary*, and the number of arbitrary relationships between two independent worlds is potentially unlimited. In the Morse code, for example, any letter of the alphabet could be associated with countless combinations of dots and dashes, which means that a specific link between them can be realized only by selecting a small number of rules. And this is precisely what a code is*: a small set of arbitrary rules selected from a potentially unlimited number in order to ensure a specific correspondence between two independent worlds*.

This definition allows us to make experimental tests because organic codes are relationships between two worlds of organic molecules and are necessarily implemented by a third type of molecules, called *adaptors*, that build a bridge between them. The adaptors are required because there is no necessary link between the two worlds, and a fixed set of adaptors is required in order to guarantee the specificity of the correspondence. The adaptors, in short, are essential in all organic codes. They are the molecular *fingerprints* of the codes, and their presence in a biological process is a sure sign that that process is based on a code.

This gives us an *objective* criterion for discovering organic codes, and their existence in Nature is no longer a matter of speculation. It is, first and foremost, an experimental problem. More precisely, we can prove that an organic code exists, if we have three things: (1) two independents worlds of molecules, (2) a set of adaptors that create a mapping between them, and (3) the demonstration that the mapping is arbitrary because its rules can, at least in principle, be changed.

## Two Outstanding Examples

*The Genetic Code*

In protein synthesis, a sequence of nucleotides is translated into a sequence of amino acids, and it has been shown that there is no necessary link between nucleotides and amino acids. These molecules belong to two independent worlds, and a bridge between them is realized by a third type of molecules, called *transfer-RNAs*, that act as adaptors and perform two distinct operations: at one site they recognize groups of three nucleotides, called *codons*, and at another site they receive amino acids from enzymes called *aminoacyl-tRNA-synthetases*. The key point is that any codon can, in principle, be associated with any amino acid. Hou and Schimmel (1988), for example, introduced two extra nucleotides in a tRNA and found that the resulting tRNA was recognized by a different synthetase and was therefore carrying a different amino acid. The number of possible connections between codons and amino acids, in other words, is potentially unlimited, and only the selection of a small fixed set of adaptors can ensure a specific mapping. This is *the genetic code*: a fixed set of rules of correspondence between codons and amino acids that are implemented by adaptors. In protein synthesis, in conclusion, we find all the three essential components of a code: (1) two independents worlds of molecules (nucleotides and amino acids), (2) a set of adaptors that create a mapping between them, and (3) the proof that the mapping is arbitrary because its rules can be changed.

*The Signal Transduction Codes*

Signal transduction is the process by which cells transform the signals from the environment, called *first messengers*, into internal signals, called *second messengers*. First and second messengers belong to two independent worlds because there are literally hundreds of first messengers (hormones, growth factors, neurotransmitters, etc.) but only four great families of second messengers (cyclic AMP, calcium ions, diacylglycerol and inositol trisphosphate) (Alberts et al. 2007). The crucial point is that the molecules that perform signal transduction are true adaptors. They consist of three subunits: a *receptor* for the first messengers, an *amplifier* for the second messengers, and a *mediator* in between (Berridge 1985). This allows the transduction complex to perform two independent recognition processes, one for the first messenger and the other for the second messenger. Laboratory experiments have proved that any first messenger can be associated with any second messenger, which means that there is a potentially unlimited number of arbitrary connections between them. In signal transduction, in short, we find all three essential components of a code: (1) two independents worlds of molecules (first messengers and second messengers), (2) a set of adaptors that create a mapping between them, and (3) the proof that the mapping is arbitrary because its rules can be experimentally changed.

## Organic Codes and Macroevolution

In addition to the genetic code and the signal transduction codes, various other organic codes have come to light, and it is likely that more will be discovered in the future. The existence of many organic codes in Nature is therefore an *experimental* fact—let us never forget this—but also more than that. It is one of those facts that have extraordinary implications.

The data from molecular biology have revealed that all cells descend from three primary kingdoms, or domains, that Carl Woese (1987, 2000) called *Archaea*, *Bacteria* and *Eucarya*. In order to understand the evolution of the first cells we need to keep in mind that bacteria appeared very early on our planet and some of them have remained substantially the same ever since. This is dramatically illustrated by the fact that modern stromatolites built by cyanobacteria are virtually identical to the 3.4 and the 1.8 billion-year-old stromatolites that have been found in the fossil record (Schopf 1999; Knoll 2003). The most primitive bacteria, in other words, already had the main characteristics of their modern descendants, whereas other primordial cells went through profound evolutionary changes. This tells us that the first cells had two evolutionary strategies in front of them, one based on increasing simplification, or *streamlining*, and one based on increasing complexity.

The cells that adopted a streamlining strategy got rid of all unnecessary components, including the ability to evolve new organic codes and have remained substantially the same ever since, whereas other cells conserved the potential to explore the coding space and have became increasingly complex.

The cells that did not evolve new organic codes became *bacteria* (*archaebacteria* and *eubacteria*) and have never fundamentally changed their cellular organization. The cells that evolved new codes, such as splicing codes, cytoskeleton codes, compartment codes, histone code and so on, became *eukarya* and have generated increasingly complex cellular structures.

We realize in this way that there is a close relationship between the appearance of new organic codes, and the appearance of genuine novelties in evolution, and we can easily understand why. The reason is that a new code brings into existence something that has never existed before because it creates arbitrary associations and generates relationships that are not determined by physical necessity.

## A New Science of Life

It is an experiment fact that the organic codes have been highly conserved in evolution, and this has revealed a totally unexpected side of life. Before the origin of the genetic code, the common ancestor was engaged in evolving coding rules and was therefore a *code exploring system*. After the origin of the code, however, no other modification in coding rules was allowed and the cell became a *code conservation system*. Another part of the ancestral cells, however, maintained the potential to evolve the rules of different codes and behaved as new *code exploring*, or *code generating, systems*. In the early *Eukarya*, for example, the cells had *a code conservation part* for the genetic code, but also a *code exploring part* for the splicing code, and this tells us something important about life.

The origin of the first cells was based on the ability of the ancestral systems to *generate* the rules of the genetic code, and the subsequent evolution of the cells was based on two complementary processes: one was the *generation* of new organic codes and the other was the *conservation* of the existing ones. Taken together, these two processes are referred to as *codepoiesis* (Barbieri 2012), a phenomenon that accounts for the two most important events that took place in evolution. The ability to create coding rules accounts for the origin of the genetic code and of all the other codes that followed. The ability of the cell to conserve its own codes accounts for the fact that the organic codes are the sole entities that have been perpetuated in evolution. They are the great invariants of life, the sole entities that have been conserved while everything else has changed.

Another outstanding implication of the existence of organic codes in Nature comes from the fact that any code involves *meaning* and we need therefore to introduce in biology, *with the standard methods of science*, not only the concept of biological information but also that of biological meaning.

The study on the organic codes, in conclusion, is bringing to light new mechanisms that operated in the history of life and new fundamental concepts. It is an entirely new field of research, the exploration of a vast and still largely unexplored dimension of the living world, the real new frontier of biology.

## About This Book

The first part of the book (Chaps. 1, 2 and 3) deals with the paradigms of biology, and in particular with the controversy between the chemical paradigm (the idea that *life is chemistry*) and the information paradigm (the idea that *life is chemistry+information*). Here it is shown that the copying of the genes and the coding of proteins are equally fundamental processes, and this leads to a third theoretical framework that is referred to as the 'code paradigm' (the idea that *life is chemistry+information+codes*). It is shown, furthermore, that genetic information and coding rules are neither quantities nor qualities but a new type of fundamental observables that are referred to as *nominable* entities. Finally, the experimental basis of the code paradigm is described by illustrating the major organic codes that have been discovered so far.

The second part of the book (Chaps. 4, 5, 6, 7, and 8) is about the role that the organic codes have played in the history of life. It is shown that the genetic code was a precondition for the origin of the first cells, the signal transduction codes divided the first cells into three primary kingdoms (*Archaea*, *Bacteria* and *Eukarya*), the splicing codes were instrumental to the origin of the nucleus, the histone code provided a new regulation system in eukaryotic genomes, and the cytoskeleton codes allowed the *Eukarya* to perform internal movements, including those of mitosis and meiosis. It is shown, furthermore, that organic codes had a key role in the transitions to multicellular life, in particular in the origin of animals, the origin of mind and the origin of language. The great events of macroevolution, in other words, were associated with the origin of new organic codes, and this gives us not only a new description but also a completely new understanding of the history of life.

The third part of the book (Chaps. 9 and 10) is about the attempts to build a theoretical framework that is based on both biological information (genetic sequences) and biological meaning (coding rules). More precisely, Chap. 9 is dedicated to *Biosemiotics*, the project launched by Thomas Sebeok in the 1980s and 1990s where the semiotics of Peirce is taken as a new '*paradigm for biology*' and all biological concepts are reformulated in Peircean terms. Chapter 10 is dedicated instead to Code Biology, the research field that studies all codes of life with the standard methods of science and where we learn *from experiments*, not from *ad hoc* definitions, what the semiotic properties of Nature actually are.

# Contents

**Part I The Paradigms of Biology**

**1 Chemistry Versus Information** ........ 3
1.1 The Chemical Paradigm ........ 4
1.2 The Information Paradigm ........ 5
1.3 Shannon's Information Theory ........ 6
1.4 Sequences and Specificity ........ 7
1.5 The Ontological Claim of the Information Paradigm ........ 9
1.6 The Ontological Claim of the Chemical Paradigm ........ 10
1.7 The Idea That *'Life Is Artifact-Making'* ........ 11
1.8 The Origin of Linear and Digital Sequences ........ 12
1.9 A Useful Metaphor ........ 13
1.10 The Issue of Meaning ........ 14
1.11 What Is Mechanism? ........ 16

**2 The Code Paradigm** ........ 19
2.1 Schrödinger's Prophecy ........ 20
2.2 The 'Special Constraints' Solution ........ 21
2.3 The New Observables ........ 22
2.4 Names and 'Nominable' Entities ........ 23
2.5 Organic Information ........ 24
2.6 Organic Meaning ........ 26
2.7 The Discovery of New Worlds ........ 27
2.8 Life and Semiosis ........ 28
2.9 The Code Model of Semiosis ........ 29
2.10 The Defining Feature of Signs and Meanings ........ 30
2.11 The Sequences of Genes and Proteins ........ 31
2.12 Two Types of Organic Signs ........ 32
2.13 A New Beginning ........ 33

3 **A Gallery of Organic Codes** ... 35
3.1 The Apparatus of Protein Synthesis ... 36
3.2 The Genetic Code ... 37
3.3 Stereochemistry and Arbitrariness ... 39
3.4 The Splicing Codes ... 40
3.5 The Metabolic Code ... 41
3.6 The Signal Transduction Codes ... 43
3.7 The Signal Integration Codes ... 44
3.8 The Histone Code ... 45
3.9 Is the "Histone Code" an Organic Code? ... 47
3.10 The Tubulin Code ... 48
3.11 The Sugar Code ... 49
3.12 The Glycomic Code ... 50
3.13 The Sequence Codes ... 52
3.14 Organic Codes and Biology ... 53

**Part II Major Steps in Macroevolution**

4 **Life Before the Cell** ... 57
4.1 The Twin Problems of the Beginning ... 58
4.2 The Replication Paradigm ... 59
4.3 Ribosoids ... 60
4.4 Nucleosoids ... 61
4.5 Nucleosoid Evolution ... 62
4.6 Heterosoids ... 64
4.7 A Primitive Apparatus ... 65
4.8 Statistical Proteins ... 66
4.9 First and Last Common Ancestor ... 68
4.10 The Ancient Genetic Code ... 69
4.11 The Modern Genetic Code ... 70
4.12 Steps Towards the Cell ... 71
4.13 The Ribotype Theory of the Cell ... 73

5 **The First Three Billion Years** ... 75
5.1 The 'Stony Carpets' ... 76
5.2 The Iron Bands ... 77
5.3 The Age of the Protista ... 78
5.4 Three Primary Kingdoms ... 79
5.5 The First Cells ... 80
5.6 Two Evolutionary Strategies ... 81
5.7 The RNA Codes ... 82
5.8 The Fluid Genome ... 84
5.9 Evolving the Cytoplasm ... 85
5.10 The Cytoskeleton Codes ... 86
5.11 The Compartment Codes ... 87

5.12 The Tree and the Web of Life ........ 88
5.13 What Happened in the Precambrian? ........ 90

**6 Evolving the Embryos** ........ 93
6.1 The Rediscovery of Epigenesis ........ 94
6.2 Body Plans and Animal Phyla ........ 95
6.3 A Cascade of Inductions ........ 97
6.4 Determination and Cell Memory ........ 98
6.5 Self-Regulating Embryonic Fields ........ 99
6.6 Arranging Cells in Space ........ 101
6.7 The Unexpected from Molecular Embryology ........ 102
6.8 The Classification of Animals ........ 104
6.9 The Cambrian Explosion ........ 105
6.10 The Origin of Animals ........ 106
6.11 The *Hox* Codes ........ 107
6.12 The Codes of the Body Plans ........ 108
6.13 The Logic of the Embryos ........ 109

**7 Brain and Mind** ........ 111
7.1 Evolving the Neuron ........ 112
7.2 The Intermediate Brain ........ 113
7.3 The Instinctive Brain ........ 114
7.4 Making Mental Images ........ 115
7.5 Sensations and Perceptions ........ 116
7.6 A Universal Neural Code ........ 117
7.7 Mechanisms of Brain Development ........ 118
7.8 Codes in Brain Development ........ 120
7.9 Theories on the Brain-Mind Relationship ........ 121
7.10 The 'First-Person' Experiences ........ 122
7.11 The Code Theory of Mind ........ 124
7.12 The Interpretive Brain ........ 125
7.13 Three Brain Macroevolutions ........ 126

**8 Origins of Language** ........ 129
8.1 Chomsky's Definitions of Language ........ 130
8.2 Sebeok's Definitions of Language ........ 130
8.3 The Bone of Contention ........ 131
8.4 A Third Foundation of Language ........ 132
8.5 A Juvenile Ape ........ 133
8.6 Fetalization and Brain Wiring ........ 134
8.7 Brain Size and Language Genes ........ 135
8.8 The *Cerebra Bifida* Model ........ 136
8.9 Mind and Reality ........ 138
8.10 Steps in the Ontogeny of Mind ........ 139
8.11 Interacting Brains ........ 140
8.12 A Hidden Asymmetry ........ 141

8.13 Rules of Neural Development ... 142
8.14 Specifically Human ... 143
8.15 The Third Cognitive System ... 144
8.16 Our Rational Faculties ... 145
8.17 The Codes of Language ... 146

**Part III A New Science of Life**

**9 Biosemiotics** ... 151
9.1 Matter Controlled by Symbols ... 152
9.2 Physical Biosemiotics ... 153
9.3 Darwinian Biosemiotics ... 154
9.4 The Peirce Framework ... 155
9.5 Zoosemiotics ... 156
9.6 Peircean Biosemiotics ... 158
9.7 Hermeneutic Biosemiotics ... 159
9.8 A Book Review for Sebeok ... 160
9.9 A Pluralistic Enterprise ... 161
9.10 Organic Semiosis ... 162
9.11 Animal Semiosis ... 163
9.12 Sense and Reference ... 164
9.13 The Worlds of Semiosis ... 165
9.14 Two Types of Biosemiotics ... 166
9.15 A New Beginning ... 168

**10 Code Biology** ... 171
10.1 What Is the Cell? ... 172
10.2 The von Neumann Machine ... 173
10.3 A Machine Capable of Self-Assembly ... 174
10.4 The Concept of Codepoiesis ... 175
10.5 Extensions of the Modern Synthesis ... 176
10.6 Mechanisms of Evolution ... 177
10.7 Copying and Coding ... 178
10.8 Codes as Constraints ... 179
10.9 Major Steps in Macroevolution ... 180
10.10 Characteristics of the Organic Codes ... 182
10.11 Different Types of Codes ... 183
10.12 Unpredictable Features ... 184
10.13 The Three Worlds of Life ... 185
10.14 Code Biology ... 186
10.15 There Was a Time ... 188

**References** .......... 191

**Author Index** .......... 207

**Subject Index** .......... 213

# Part I
# The Paradigms of Biology

# Chapter 1
# Chemistry Versus Information

From time immemorial it has been taken for granted that life is fundamentally different from matter, but in the last few centuries this belief has been seriously challenged by the view that '*life is chemistry*'. The idea that life evolved naturally on the primitive Earth suggests that the first cells came into being from previous chemical systems by spontaneous chemical reactions, and this is equivalent to saying that there is no *fundamental* divide between life and matter.

This is the *chemical paradigm*, a view that is very popular today and that is often considered in agreement with the Darwinian paradigm, but this is not the case. The reason is that natural selection, the cornerstone of Darwinian evolution, does not exist in inanimate matter. In the 1950s and 1960s, furthermore, molecular biology has uncovered two fundamental components of life – biological information and the genetic code – that are totally absent in the inorganic world, which means that information is present only in living systems, that chemistry alone is not enough and that a deep divide does exist between life and matter. This is the *information paradigm*, the idea that '*life is chemistry* + *information*'.

Ernst Mayr, one of the architects of the Modern Synthesis, has been one of the most outspoken supporters of the view that life is *fundamentally* different from inanimate matter. In *The Growth of Biological Thought* (1982), he made this point in no uncertain terms:

> ... The discovery of the genetic code was a breakthrough of the first order. It showed why organisms are fundamentally different from any kind of nonliving material. There is nothing in the inanimate world that has a genetic program which stores information with a history of three thousand million years! (p. 124)

> ... Except for the twilight zone of the origin of life, the possession of a genetic program provides for an absolute difference between organisms and inanimate matter. (p. 56)

The discoveries of molecular biology, in short, appear in contrast with the chemical paradigm, and this raises formidable problems. On the one hand it is an experimental fact that natural selection, biological information and the genetic code do not exist in inanimate matter. On the other hand, we seem unable to accept that

© Springer International Publishing Switzerland 2015

M. Barbieri, *Code Biology*, DOI 10.1007/978-3-319-14535-8_1

life evolved from inanimate matter and yet it is fundamentally different from it. How can something give origin to something fundamentally different from itself? How could the physical world produce life if there is a discontinuity between them?

The aim of this chapter is to show that a logical answer to these questions does exist, but it does not come from the chemical paradigm or from the information paradigm. It comes instead from a new idea on the nature of life that becomes the starting point of a third paradigm.

## 1.1 The Chemical Paradigm

Ever since the scientific revolution, physics has been the 'queen' of the sciences, and biologists have been split into opposite camps, one in favour and one against adopting its method, an approach which has become known as *mechanism*. In biology, the first version of mechanism was the Cartesian doctrine that "*the body is a machine*" and that the clock is its model: *"A healthy man is like a well functioning clock, and an ill man is like a clock that needs repairing"* (Descartes 1637).

The mechanical concept of nature spread very quickly in seventeenth century Europe, but not without conflict. Opposition came particularly from a new science that was slowly emerging from alchemy and that regarded the human body essentially as a seat of chemical reactions. The heirs of the alchemists were determined to leave magic behind but had no intention of accepting the 'mechanical' view of nature, and one of chemistry's founding fathers, Georg Ernst Stahl (1659–1731), launched an open challenge to mechanism. He claimed that organisms cannot be machines because what is taking place inside them are not movements of wheels, belts and pulleys but real transmutations of substances.

The arguments of the chemists did have an impact, and eventually forced mechanists to change their model. In the course of the eighteenth century, the view that organisms are *mechanical machines*, gradually turned into the idea that they are *chemical machines.* This change went hand in hand with the development of the *steam engine*, and that machine became the new model of biology. In the nineteenth century, furthermore, the study of the steam engine was pushed all the way up to the highest level of theoretical formalism, and culminated with the discovery of the first two laws of thermodynamics. The result is that living systems came to be seen as *thermodynamic machines*, i.e., as chemical machines that must be continuously active in order to obey the laws of thermodynamics.

The old opposition between physics and chemistry came to an end, and the two sciences gave origin to a unified approach that is often referred to as the *chemical paradigm*, the idea that '*life is chemistry*' or, more precisely, '*an extremely complex form of chemistry*'. This is equivalent to saying that all biological processes are chemical transformations of matter and energy and are completely described, at least in principle, by relationships between physical quantities.

The chemical paradigm has underlined time and again – against all forms of vitalism – that living systems are subject to the laws of thermodynamics, but it is by no means confined to this principle. It is a paradigm which has steadily grown by the addition of new developments. The non-equilibrium thermodynamics of Ilyia Prigogine, for example, the phase-transitions of Stuart Kauffman, chaos theory and complexity theory, are all new research fields that rightly belong in the framework of the chemical paradigm.

The same is true for the idea that physical forces and mathematical principles are life-generating entities, a recurrent theme in the history of science, from Goethe and D'Arcy Thomson to Renè Thom and the recent field of *Systems Biology*. The chemical paradigm, in short, is the view that the laws of physics and chemistry, together with the principles of logic and mathematics, are all that we need to account for the presence of life in the universe.

## 1.2 The Information Paradigm

At the beginning of the twentieth century, the rediscovery of the laws of Mendel led Wilhelm Johannsen to make a sharp distinction between the visible part of an organism (the *phenotype*) and the invisible part that carries its hereditary information (the *genotype*). Johannsen (1909) proposed that every living being is a dual entity, a synthesis of two complementary realities. This idea was largely ignored, at first, but a few decades later the computer made it immediately comprehensible. The *phenotype-genotype* duality was a *hardware-software* distinction, and became the prototype description of any organism. The model of the living system changed again and became the *computer*.

In 1953, James Watson and Francis Crick pointed out that the sequence of nucleotides represents the *information* carried by a gene (Watson and Crick 1953). A few years later, the mechanism of protein synthesis was discovered and it was found that the sequence of nucleotides in genes determines the sequence of amino acids in proteins, with a process that amounts to a transfer of linear information from genes to proteins (Crick 1957). This led to the idea that *biological information* is the specific sequence in which the subunits of a molecular polymer are arranged.

These discoveries gave origin to the *information paradigm*, the second great theoretical framework of modern biology. It is the idea that living systems are *information-processing machines*, and that life is based not only on chemistry (energy and matter) but also, and above all, on *information* (Maynard Smith 2000). In this framework, chemistry accounts for the hardware of living systems whereas information provides the software, and the view that '*life is chemistry*' was replaced by the idea that '*life is chemistry + information*'.

This, in turn, led to the concept of the 'genetic program', the idea that the genome is for the cell what a program is for a computer. The logical separation that exists between program and machine implies that something similar exists between the genome and the cell, and such a biological separation has in fact been documented by an outstanding number of experimental results. Many genes, for example, have been transplanted from one type of cells to another and have turned out to be fully functional in the new cells (Danchin 2009). Many bacteria now produce human proteins, and the very existence of viruses can be explained by the transmission of independent genetic strings, thus confirming that genes are separable from the cell machine. It has even been possible to transplant an entire genome from one species to another, thus proving that genomes do have a substantial degree of autonomy (Lartigue et al. 2007).

This informational view of life has been fully accepted by the Modern Synthesis because the concept of information goes hand in hand with heredity and natural selection. Heredity is the transmission of genetic information from one generation to the next, the short-term result of molecular copying, whereas the long-term repetition of copying is inevitably accompanied by errors, and in a world of limited resources not all copies can survive and a selection is bound to take place.

Today, in other words, heredity and natural selection are squarely based on biological information, and the information paradigm has become, to all effects, the core of the Modern Synthesis, a view of life which is in conflict with the chemical paradigm, because information, heredity and natural selection simply do not exist in the world of chemistry.

## 1.3 Shannon's Information Theory

The concept of information has been introduced in science in two very different ways. In biology, as we have seen, it was introduced by Watson and Crick in 1953 and was identified with the specific sequence of nucleotides in a DNA molecule. In engineering, the concept was introduced by Claude Shannon in 1948 with an entropy-like formula, and for this reason it is referred to as *statistical information.*

Shannon conceived information as an entity that is generated whenever uncertainty is reduced. When we receive a message, for example, we pass from a state of ignorance to one of knowledge, as we do when we toss a coin and pass from a state of equal probabilities to one of certainty. This is why Shannon proposed to *measure* the information quantity of a message with the probability of choosing that message from the set of all potential alternatives (Shannon 1948).

Shannon was particularly interested in telephone transmissions and described any communication system as a combination of a *source* (that produces signals), a *destination* (that receives them) and a *channel* in between. The goal of communication engineering is to ensure that a message transmitted from the source is reproduced at destination with the minimal amount of distortion, whatever is the length, the structure and the *meaning* of the message. The engineering problem is concerned therefore with the reduction, and possibly the elimination, of the noise that inevitably affects the transmission phase, and to this purpose the message is submitted to two anti-noise operations that are known as *source coding* and *channel coding*.

Source coding consists in reducing the length of a message by getting rid of all unnecessary features (or *redundancies*) in order to minimize the exposition to noise. Channel coding is instead an anti-noise operation that is achieved not by reducing but by increasing the length of a message, and is obtained by introducing a variety of noise-detecting sets of characters that are known as *error-correcting-codes* (Battail 2007). We can get an intuitive idea of channel coding by comparing it to the familiar operation that we resort to in noisy situations when we say, for example, "Alpha, Bravo, Charlie, etc." in place of "a, b, c, etc."

The extraordinary result obtained by Shannon was a set of theorems which prove that reliable, error-free, communication is possible even in unreliable channels. Within ample and well defined limits, noise can be reduced by as much as we want to, a result that paved the way to the tremendous expansion of the communication technologies, especially after the introduction of a powerful family of error-correcting-codes that are known as *Turbo codes* (Battail 2008).

It must be underlined, however, that the noise reducing operations of source coding and channel coding are unproblematic when the messages consist of discrete characters, or *digits*, but it is virtually impossible to remove noise from *analog* signals, and it is for this reason that industry and media have steadily moved from analog to digital technologies.

The new field of research established by Shannon has become known as *Information Theory*, and in engineering it has been extremely successful but in biology its impact has been rather limited. The reasons for this are still far from clear, and it may well be that the potential of Shannon's theory for the life sciences has not yet fully come to light and remains a challenge for the future (Battail 2007, 2010, 2014).

## 1.4 Sequences and Specificity

There is a clear similarity between a sequence of nucleotides in a gene and a sequence of letters in a word, and in both cases we say that sequences carry information: hereditary information in genes and syntactic information in language. This concept, furthermore, can be generalized to many other sequences. A painting or a photograph, for example, can be digitized and represented by a matrix of pixels arranged in rows and columns, but the rows can also be arranged one after the other in a line and give origin to a one-dimensional string, i.e., to a sequence of pixels. The same can be done with sounds and music, first by digitizing them and then by arranging their elements in a linear order. Finally, we can represent letters, numbers and many other symbols with the characters of computer language, and any sequence can be described as a sequence of *bytes*. More precisely, we can represent in computer language any configuration of abstract or concrete objects that is (1) linear, (2) digital and (3) finite.

We obtain in this way a first definition: *a sequence is any collection of a finite number of digital objects that are arranged in a linear order*. The fact that the objects (nucleotides, letters, pixels, musical notes etc.) are arranged on a line in a precise way, and not at random, means that a sequence necessarily has a unique order, and it is exactly this order that biologists call *information*. More precisely it is called *sequence information*, because all sequences have it, and often it is also called *specificity* to underline that its defining feature is the specific order of its components.

A sequence defined in this way has two outstanding characteristics. The first is that specificity has nothing to do with meaning. The sequence information of the word 'ape', for example, is the same in all languages, but in English it means 'tailless monkey', whereas in Italian it means 'bee' and in French it has no meaning. Specificity, or sequence information, in other words, is a syntactic entity, not a semantic one, and for this reason it is often referred to as *syntactic information*.

The second outstanding characteristics is that specificity *cannot be measured*. It can only be identified by *naming* its components in their natural order. This is something that biologists, and in particular geneticists, tend to ignore. The usual reaction to the statement that genetic sequences cannot be measured is "What are you talking about? We measure them everyday – it's called genotyping".

In order to clarify this point we need to keep in mind that there are two features in a sequence that can be measured. The first is the *size*, or the *length*, of the sequence, a quantity that represents the total number of its characters, or the number of bytes that we need to store it in a computer's memory.

The second is the *relative distance* that exists between two sequences, a quantity that measures the degree of *compatibility*, or *relatedness*, that exists between them. Genotyping, for example, is the technique that measures the number of nucleotides that two genetic samples have in common. The same technique, applied to literature, allows us to measure the relative distance that exists between two books, or two languages, by counting the number of words that they have in common. The relative distance that exists, for example, between *Hamlet* and *Macbeth*, gives us a measure of their relatedness, but has nothing to do with the specific order of their words, i.e., with their specificity, and even less with their meanings.

Any sequence, in conclusion, is characterized by four distinct entities, two of which can be measured whereas the other two are not measurable: (1) the *length* of a sequence can be measured (in bytes), (2) the relative *distance* of a sequence from another sequence can be measured, (3) the *specificity* of a sequence cannot be measured, and (4) the *meaning* of a sequence cannot be measured (and may not exist).

The fact that in *all* sequences – organic or linguistic – specificity can only be named, inevitably raises a question: if specificity, or sequence information, cannot be measured, what is it? What does it represent in Nature? This is an extremely important point, because, as we will see, the difference between the Chemical paradigm and the Information paradigm is precisely the fact that they give different answers to this question. The *nature* of sequence information, in other words, is the key issue that divides the two present paradigms of biology.

## 1.5 The Ontological Claim of the Information Paradigm

The discovery of biological information was the event that transformed biochemistry into molecular biology, and the paradigm that '*life is chemistry*' into the new paradigm that '*life is chemistry-plus-information*'. This idea implies that information is *ontologically* different from chemistry, but can we prove it? Ontology is the study of being and saying what an entity is amounts to defining it. Ontology, in short, is concerned with the definition of entities at the most basic level.

The ontological claim of the Information paradigm is that life is *fundamentally* different from matter because there is an *ontological* difference between information and chemistry. Unfortunately, however, the proponents of the Information paradigm have never *proved* this claim. Ernst Mayr, as we have seen, has repeatedly stated that the existence of the genetic code is enough to prove that life in fundamentally different from chemistry, but has not been able to say why.

Perhaps the strongest criticism of the chemical paradigm has come from the Information Theory camp, and in particular from Hubert Yockey, one of the organizers of the first congress dedicated to the introduction of Shannon's information in biology (Yockey et al. 1958). In a long series of articles and books, Yockey (1974, 2000, 2005) has underlined that heredity is transmitted by factors that are "segregated, linear and digital" whereas the compounds of chemistry are "blended, three-dimensional and analog".

> Chemical reactions in non-living systems are not controlled by a message. If the genetic processes were purely chemical, the law of mass action and thermodynamics would govern the placement of amino acids in the protein sequences according to their concentrations ... There is nothing in the physico-chemical world that remotely resembles reactions being determined by a sequence and codes between sequences (Yockey 2000).

Yockey has tirelessly pointed out that no amount of chemical evolution can cross the barrier that divides the *analog world of chemistry* from the *digital world of life*, and concluded from this that the origin of life cannot have been the result of chemical evolution. This is therefore, according to Yockey, what divides life from matter: information is ontologically different from chemistry because linear and digital sequences cannot be generated by the analog reactions of chemistry.

At this point one would expect to hear from Yockey how did linear and digital sequences appear on Earth, but he did not face that issue. He claimed instead that the origin of life is *unknowable*, in the same sense that there are propositions of logic that are *undecidable*. The problem, with this argument, is that the existence of undecidable propositions has been *proven* in logic, whereas the conclusion that the origin of life is unknowable is just an assumption. It may be a legitimate assumption, in principle, but in no way it is comparable to Gödel's theorem and certainly it does not carry the same weight.

It is important however to recognize that Yockey's distinction between analog and digital entities cannot be ignored. He was absolutely right in saying that the spontaneous reactions of chemistry cannot produce molecules with linear and digital sequences, and this is indeed the crucial problem that must be faced by any scientific theory on the origin of life. The information paradigm has not solved this problem and it is for this reason that it has not been able to prove its ontological claim.

## 1.6 The Ontological Claim of the Chemical Paradigm

The view that *'life is chemistry'* was proposed for the first time by Jan Baptist van Helmont (1648), and has been re-proposed countless times ever since. One of the most recent formulations has been given by Günther Wächtershäuser (1997) in these terms "If we could ever trace the historic process backwards far enough in time, we would wind up with an origin of life in purely chemical processes".

He added that "The science of chemistry, however, is an ahistoric science striving for universal laws… so this is the challenge of the origin of life: to reduce the historic process of biological evolution to a universal chemical law of evolution". The difficulty of this task, he pointed out, is due to the fact that "Chemistry is mechanistic and history teleological, and the life sciences are the arena where mechanistic explanations and teleological understanding come into close encounter."

Wächtershäuser claimed that "information is a *teleological* concept", and gave a specific example of the conflict between mechanism and teleology: "On the level of nucleic acid sequences it is quite convenient to use the information metaphor … and apply teleological notions such as 'function' or 'information' … but in the course of the process of retrodiction the teleological notions, whence we started, fade away. And what remains is purely chemical mechanism". This amounts to saying that biological information, the most basic concept of molecular biology, does not *really* belong to science.

This is the ontological claim of the Chemical paradigm, the idea that all natural processes are completely described, in principle, by physical quantities. This view is also known as *physicalism*, and it is based on the fact that biological information, or biological *specificity*, is not a physical quantity. So, what is it? A similar problem arises with the genetic code. The rules of a code cannot be measured and cannot be reduced to physical quantities. So what are they?

According to physicalism, biological information and the genetic code are mere *metaphors*. They are linguistic expressions that we use as shortcuts in order to avoid repeating every time all the details of long chains of chemical reactions. But behind those terms there are only chemical reactions and nothing else. They are like those computer programs that allow us to write our instructions in English, thus saving us the trouble to write them in the binary digits of the machine language. Ultimately, however, there are only binary digits in the machine language of the computer, and in the same way, it is argued, there are only physical quantities at the most fundamental level of Nature.

This conclusion, known as the *physicalist thesis*, has been proposed in various ways by a number of scientists and philosophers (Chargaff 1963; Sarkar 1996, 2000; Mahner and Bunge 1997; Griffith and Knight 1998; Griffith 2001; Boniolo 2003), and is equivalent to the thesis that *'life is chemistry'*.

This is one of the most deeply dividing issues of modern science. Many biologists are convinced that biological information and the genetic code are real and fundamental components of life, but physicalists insist that they are real only in a very superficial sense and that there is nothing fundamental about them because they must be reducible, in principle, to physical quantities. An additional complication is the fact that many physicists are *not* physicalists, and fully accept the idea that a deep divide does exist between life and matter.

## 1.7 The Idea That *'Life Is Artifact-Making'*

According to the chemical paradigm, the first cells evolved from chemical systems by spontaneous chemical reactions that can be fully described, in principle, by physical quantities. No other entities are required to explain the origin of life by chemical evolution, and this is why physicalism concludes that biological information and the genetic code are purely metaphorical terms.

It must be underlined that the physicalist thesis would be absolutely correct if genes and proteins were spontaneous molecules because there is no doubt that all spontaneous reactions are completely accounted for by physical quantities. This, however, is precisely the point that molecular biology has proved wrong. Genes and proteins are *not* produced by spontaneous processes in living systems. They are produced by molecular machines that physically stick their subunits together according to sequences and coding rules and are therefore *manufactured molecules*, i.e., *molecular artifacts*. This in turn means that all biological structures are manufactured, and therefore that the whole of life is *artifact-making* (Barbieri 2004, 2006a, 2008). This conclusion may appear paradoxical, at first, but let us take a closer look.

Biochemistry has long argued that virtually all reactions that take place in living systems are catalyzed processes, but molecular biology has discovered that the production of genes and proteins requires not only catalysts but also *templates*. The catalysts join the subunits together by chemical bonds, and the templates provide the *order* in which the subunits are assembled. It is precisely that order that determines biological specificity, the most important characteristic of life, and that order comes from a molecule that is *outside* the assembled molecule.

This is precisely the characteristic that divides spontaneous objects from artifacts. In spontaneous and in catalyzed processes, the order of the components comes *from within* the molecules, i.e., is determined by *internal* factors, whereas in genes and proteins it comes *from without*, from an *external* template.

The difference between spontaneous and manufactured objects, in short, does not exist only at the macroscopic level of culture. It also exists at the molecular level, because it is an *experimental fact* that genes and proteins are *template-dependent* molecules, and this means that they are manufactured molecules and therefore molecular artifacts.

Let us now look at the difference between the processes that manufacture genes and proteins. They both require catalysts and templates, but in addition to that proteins also require a set of coding rules (in the form of molecular adaptors). This is because genes are nucleic acids that are formed by copying a template, whereas proteins cannot be copied. Their order must still come from nucleic acids (because only these molecules can be inherited) but a sequence of nucleic acid has to be translated into a sequence of amino acids and this is achieved, in protein synthesis, by the rules of the genetic code.

We realize in this way that there are two distinct processes at the basis of life: the *copying* of genes and the *coding* of proteins. Genes are manufactured by molecular machines that can be referred to as *copymakers* and proteins by molecular machines that can be called *codemakers*. Copying and coding, on the other hand, are both artifact-making processes and life as we know it requires both of them. We can truly say therefore that *life is artifact-making*, or, more precisely, that *life is artifact-making by copying and coding*.

This makes us realize that the physicalist thesis is wrong because it is only spontaneous processes that are completely described by physical quantities. Manufacturing processes require additional entities, like sequences and coding rules, and these are not physical quantities, because they cannot be measured, but are nevertheless absolutely essential to the description of living systems.

## 1.8 The Origin of Linear and Digital Sequences

The existence of linear and digital sequences in life is a fact, an experimental fact, and all biologists acknowledge it. It is equally a fact that linear and digital sequences that direct the synthesis of molecules do not exist in the inanimate world, so it is beyond dispute that a divide does exist between life and matter. It is the divide between the analogue world of chemistry and the digital world of life, and it is not a fiction. The problem is the origin of that divide, not its existence.

Hubert Yockey has underlined that spontaneous reactions cannot produce a living cell, and that, let us repeat it, is formally correct. The real answer to Yockey is not a denial of this point, but the argument that it does not apply to living cells because these are *not* produced by spontaneous reactions. The evidence shows that genes and proteins are manufactured by molecular machines in all present cells, and the most logical conclusion we can draw is that this has been true also in all cells of the past, including the first cells.

Yockey's critique of chemical evolution is justified only if we assume that chemical evolution was but a sequence of *spontaneous* reactions, because linear, digital and specific properties do not exist in spontaneous processes. But they do exist in all manufacturing processes, including those that are produced by molecular machines at the molecular level. The answer to Yockey's argument, in short, is that genes and proteins are molecular artifacts, that life itself is artifact-making (Barbieri 2003, 2006a).

When a copymaker scans a nucleic acid and makes a copy of that molecule, what happens is precisely an operation that brings into existence a linear and digital copy of a pre-existing molecule. It was molecular copying – the simplest form of artifact-making – that started manufacturing biological objects and set in motion the odyssey of life on the primitive Earth.

What is particularly important, to our purposes, is that the concept of artifact-making explains how it is possible that life evolved from inanimate matter and yet it is fundamentally different from it. The divide between life and matter is real because matter is made of spontaneous objects whereas life is made of manufactured objects. The idea that life is artifact-making, in short, is the only logical alternative to the chemical paradigm, and allows us to study the origin of life as a natural phenomenon that was brought into existence by the evolution of molecular machines.

## 1.9 A Useful Metaphor

We find it difficult to accept that life evolved from matter and, at the same time, that it is fundamentally different from it. How can something give origin to something fundamentally different from itself? The way out of this dilemma, as we have seen, is the idea that *life is artifact-making*, i.e., that the key molecules of life did not arise spontaneously but were brought into existence by molecular machines. This idea, however, does not seem intuitively appealing, so it may be useful to illustrate it with a metaphor. It is a sort of cartoon, if you like, but if used consistently it is as rigorous as a technical argument.

The metaphor consists in saying that all spontaneous molecules are 'grey' (all shades of grey between white and black), whereas all manufactured molecules are 'coloured' (all colours of the rainbow). With this terminology, the concept that life is artifact-making amounts to saying that the world of life is coloured whereas the world of inanimate matter is grey, and this gives us a new way of formulating the problem of the origins. Earth was a lifeless planet, at the beginning, and all its molecules were grey, so how did coloured molecules appear out of grey matter?

Spontaneous genes and spontaneous proteins did appear on the primitive Earth but they did not evolve into the first cells, because they did not have biological specificity. They gave origin to *molecular machines* and it was these machines and their products that evolved into the first cells. The simplest molecular machines that could appear spontaneously on the primitive Earth were molecules that could stick monomers together, first at random (*bondmakers*) and then in the order provided by a template (*copymakers*).

These molecules started manufacturing polymers such as polypeptides, polynucleotides and polysaccharides, and had the potential to produce them indefinitely, thus increasing dramatically their presence on the primitive Earth. The origin of proteins, on the other hand, was a much more complex affair, because proteins cannot be copied and their reproduction required machines that developed a *code* and which can therefore be referred to as *codemakers*. The evolution of the molecular machines, in short, started with bondmakers, went on to copymakers and finally gave origin to codemakers.

If we translate all this in the terminology of grey and coloured molecules, we can say that the first molecular machines were grey (because they appeared spontaneously) and that they started producing coloured molecules (because manufactured molecules are coloured). The first molecular machines were therefore a special type of grey molecules, and we may call them *silver* molecules. The machines that came after them, however, could incorporate also coloured molecules, and eventually these replaced all grey elements. The silver molecular machines evolved into coloured machines and we can illustrate this transformation by saying that they became *golden* molecular machines. At this stage, the divide between life and matter became complete, because all the components of life, molecules and molecular machines, were coloured (*manufactured*), whereas all components of inanimate matter were grey (*spontaneous*).

The transition from spontaneous to manufactured molecules is documented by many examples, but here it will be enough to mention one that is particularly illuminating. The amino acids that can be synthesized from inorganic molecules are *less than half* of the 20 canonical ones, and are referred to as *primary* (or *precursor*) amino acids because they serve as starting points for the synthesis of the remaining ones, the so-called *secondary* (or *product*) amino acids. This implies that only primary amino acids appeared spontaneously on the primitive Earth (Wong and Bronskill 1979; Wong 1981; Higgs and Pudritz 2007). What happened afterwards, is that some primitive systems evolved the ability to manufacture secondary amino acids and eventually also the primary ones. In the history of life, in other words, the sequence of appearance was: (1) spontaneous, (2) spontaneous-plus-manufactured and (3) manufactured amino acids. Which in our metaphor is equivalent to saying that the first amino acids were all gray, then they were both gray and coloured and finally all amino acids became coloured molecules.

## 1.10 The Issue of Meaning

The double helix and the genetic code have been two of the major scientific discoveries of all times and yet, surprisingly, the *old regime* has not been deposed. The majority view, today, is still the idea that life is an extremely complex form of chemistry that evolved spontaneously from primitive chemical systems. This view is based on the physicalist thesis that all biological processes are completely described, in principle, by physical quantities, which means that no other fundamental entity is required to account for them. Entities like genetic information and coding rules are regarded as metaphorical terms that people use simply because they are intuitively appealing. In such a framework, the revolution of molecular biology amounts to little more than the introduction of fancy names into the solid body of biochemistry.

This is the great paradox of modern biology. On the one hand, genetic information and the genetic code have become the bread and butter of biological research, and on the other hand we are told that they are mere linguistic decorations. The paradox is due to the fact that the information paradigm has *claimed* that genetic information is a new fundamental entity, but has not been able to say *why*. The stalemate between the two present paradigms of biology, in other words, is due to the fact that *the information paradigm has not provided a real alternative to the chemical paradigm*.

Here we have seen that such an alternative does exist, because the physicalist thesis is valid only in spontaneous systems whereas genes and proteins are never formed by spontaneous reactions. They are invariably manufactured by molecular machines, and all manufacturing processes do not require only physical quantities but also additional entities like sequences and codes. The alternative to the view that '*life is chemistry*', in short, is the view that '*life is artifact-making*'.

Unfortunately, modern biology has accepted the concept of organic information but not the concept of organic meaning, and this is equivalent to saying that genetic information is real but the genetic code is not. More precisely, meaning has been regarded not as an entity in its own right, but as a 'qualification' of information. Rather than talking of information and meaning, many biologists are talking of "meaningful information", "semantic information", "functional information" and the like.

In a recent review entitled *Information in Biological Systems* John Collier (2008) has listed at least seven different types of information that apparently form a nested hierarchy: (1) physical information, (2) statistical information, (3) expressed information, (4) functional information, (5) meaningful information, (6) intentional information, and (7) social information.

Similar proposals have been made by many other authors with different terminologies, and there seem to be no end in sight to the proliferation of the information categories. But why does this happen? Why do we keep multiplying the types of information in order to account for properties that belong to the category of meaning? It is high time to acknowledge that information and meaning are two distinct entities and that in biology too we must face the issue of meaning.

What we need, in other words, is a new paradigm that fully accepts the implications of the discovery that life is based on copying *and* coding, and that these processes necessarily require sequences and codes. We need a paradigm where biological sequences (organic information) and coding rules (organic meaning) are *real* and *fundamental* entities of Nature, as real and fundamental as the physical quantities.

There is however a massive obstacle that apparently blocks our way on this road: introducing the concept of meaning in biology seems equivalent to reintroducing the old spectre of vitalism, and this is something that biologists simply are not prepared to consider. But is that true? Is the concept of organic meaning really incompatible with the good old-fashioned *mechanism* that lies at the heart of the scientific method?

## 1.11 What Is Mechanism?

The model of the chemical paradigm is the steam-engine whereas the model of the information paradigm is the computer. Both of them are very different from the clock-model of Descartes but they are all *mechanistic* models of life, so we need to ask ourselves '*what is mechanism?*'

One of the expressions that best catches the spirit of mechanism is John Maynard Smith's statement that *"We understand biological phenomena only when we have invented machines with similar properties"* (Maynard Smith 1986).

In fact, 'understanding' something means explaining it with a model that we are familiar with, and a machine gives us an immediate sense of familiarity. When we see it working before our eyes, we feel that we 'know' it. Actually, we do not even need to build a machine to get this feeling. A description is enough, and so a machine is often a *model*, or even an *algorithm*. One of the most famous machines of all times was built by Turing with just pencil and paper.

A model, furthermore, does not necessarily have a mathematical form. Natural selection, for example, is a mechanistic model which is entirely expressed in words. The important point is that the model has the *logic* of a machine (i.e. that it delivers the same sense of familiarity that we get from a real functioning machine). Mechanism, in short, is the view that scientific knowledge is obtained by building machine-like models of what we observe in nature. Let us briefly summarize it.

1. Mechanism is not *reductionism*, because a machine is a machine not when it is reduced to pieces but when it is put together into a working whole.
2. Mechanisms is not *determinism*, because it is more general than classical physics (quantum theory is mechanism, and so is non-equilibrium thermodynamics, chaos theory and the like).
3. Mechanism is not *physicalism*, because it is not limited to physical quantities (natural selection, the Turing machine and Godel's theorem are mechanistic models that are not based on physical quantities).
4. Finally, and most importantly, mechanism is made of models and models do not coincide with reality, a concept that Count Alfred Korzybski (1996) vividly expressed with the maxim *"the map is not the territory"*. This means that mechanism is intrinsically incomplete and continuously evolving.

Mechanism, in short, is virtually equivalent to the scientific method. The difference is that the *hypotheses* of the scientific method are replaced by *models*, i.e., by descriptions of fully functional working systems. Mechanism, in other words, is '*scientific modelling*'.

Ever since it first appearance, at the beginning of the scientific revolution, mechanism has been highly effective in accounting for particular aspects of Nature, and at the same time it has shown an extraordinary ability to change in the face of adversity. The first mechanistic model of the body was the clock-machine, then came the steam-engine-machine, and after that the computer-machine. Which amounts to saying that mechanism has introduced in biology first *mechanical energy*, then *chemical energy*, and finally *information*.

Now we face a new challenge, and once again we hear that mechanism is not enough, that we need something completely different. Which could be true, of course, but mechanism remains our best chance to find out what makes living systems tick. Mechanism may well be able to change again and introduce in biology not only the concepts of energy and information, but also the last frontier, the concept of *meaning*.

# Chapter 2
# The Code Paradigm

Today there are two major theoretical frameworks in biology. One is the 'chemical paradigm', the idea that life is an extremely complex form of chemistry. The other is the 'information paradigm', the view that life is not just '*chemistry*' but '*chemistry-plus-information*'. This implies the existence of a *fundamental* difference between information and chemistry, a conclusion that is strongly supported by the fact that information and information-based-processes like heredity and natural selection simply do not exist in the world of chemistry.

Against this conclusion, the supporters of the chemical paradigm have pointed out that there is no real difference between chemical processes and information processes because both are described by physical quantities. They *may* appear different, it is argued, but this is only because they have different degrees of complexity. According to the chemical paradigm, in other words, biological information is but a shortcut term that we use in order to avoid long descriptions of complex chemical reactions. It is an intuitively appealing concept, but it does not represent a new *ontological* entity. It is merely a derived construct, a linguistic metaphor.

The supporters of the information paradigm insist that information is a new fundamental entity of Nature, but have not been able to *prove* this point. The result is that the chemical view has not been abandoned and today biology is in the uncomfortable position of having two contrasting paradigms at its basis.

Here it is shown that a solution does exist and it is a third theoretical framework that is referred to as the *code paradigm*. The key point is that we need to introduce in biology not only the concept of information but also that of meaning because any code is based on meaning and a genetic code does exist in every cell. The third paradigm, in short, is the view that organic information and organic meaning exist in every living system because they are the inevitable results of the processes of copying and coding that produce genes and proteins. According to this view, life is '*chemistry-plus-information-plus-codes*'.

The true nature of organic information and organic meaning has for a long time eluded us because they exist only in artifacts and biologists have not yet come to terms with the idea that life is *artifact making*. This is the idea that life

© Springer International Publishing Switzerland 2015

M. Barbieri, *Code Biology*, DOI 10.1007/978-3-319-14535-8_2

arose from matter and yet it is fundamentally different from it, because inanimate matter is made of spontaneously formed objects whereas life is made of objects that are manufactured by molecular machines. It will be shown, furthermore, that the existence of information and meaning in living systems is documented by the standard procedures of science. We do not have to abandon the scientific method in order to introduce them in biology. All we need is the experimental proof that organic codes do exist in Nature.

## 2.1 Schrödinger's Prophecy

In 1944, Erwin Schrödinger wrote *"What is Life?"*, a little book that inspired generations of scientists and became a landmark in the history of molecular biology. There were two seminal ideas in that book: one was that the genetic material is an *"aperiodic crystal"*, the other was that *"the chromosomes contain a code-script for the entire organism"*.

The metaphor of the aperiodic crystal was used by Schrödinger to convey the idea that the atoms of the genetic material must be arranged in a unique pattern in every individual organism, an idea that later was referred to as *biological specificity*. The metaphor of the code-script was used to express the concept that there must be *a miniature code* in the hereditary substance, a code that Schrödinger compared to *"a Morse code with many characters"*, that was supposed to carry *"the highly complicated plan of development of the entire organism"* (Schrödinger 1944). That was the very first time that the word 'code' was associated with a biological structure and was given a role in organic life.

The existence of specificity and code at the heart of life led Schrödinger to a third seminal conclusion, an idea that he expressed in the form of a prophecy: *"Living matter, while not eluding the 'laws of physics' as established up to date, is likely to involve hitherto unknown 'other laws of physics', which, however, once they have been revealed, will form just an integral part of this science as the former"*.

Schrödinger regarded this prophecy as his greatest contribution to biology, indeed he wrote that it was *"my only motive for writing this book"*, and yet that is the one idea that even according to his strongest supporters did not stand up to scrutiny. Some 30 years later, Stent and Calendar (1978) gave up the struggle and concluded that *"No 'other laws of physics' turned up along the way. Instead, the making and breaking of hydrogen bonds seems to be all there is to understanding the workings of the hereditary substance"*.

Schrödinger's prophecy of new laws of physics appears to have been shipwrecked in a sea of hydrogen bonds, but this is true only in a superficial sense. The essence of the prophecy was the idea that the two basic features of life – specificity and the genetic code - require *new fundamental entities* of Nature that are *"hitherto unknown"*, and in that form it is still valid. The fact that Schrödinger invoked new *laws of physics* should not be allowed to obscure the substance of the prophecy, which can be expressed in this way: in order to understand life we need to discover *something fundamentally new*, something that is still not part of physical theory.

Let us turn therefore to this generalized version of Schrödinger's prophecy. He anticipated the concept of biological specificity, or biological information, and announced that there must be a 'code-script' in every living cell. Both ideas were truly prophetic at the time and both turned out to be true. That should be enough for us to reconsider the *essence* of his prophecy: is it true that we need *something fundamentally new* in order to explain biological information and the genetic code?

## 2.2 The 'Special Constraints' Solution

In the 1960s, Howard Pattee pointed out that the theory of self-replicating machines developed by John von Neumann in the 1950s throws an important light on the genetic code. Von Neumann had shown that a self-replicating system must contain a description of itself, and such a description must be categorically different from the system ('*the map is not the territory*'). The description of a system, on the other hand, is necessarily made of entities that represent, or 'stand for', its material components, and function therefore as signs or symbols. According to von Neumann, in short, a self-replicating system must be *a physical system controlled by symbols*, or, more precisely, by a program, by the rules of a code (von Neumann 1951, 1958, 1966).

This was enough, according to Pattee, to prove that every living cell is controlled by a real code, and he set out to find out how physical theory can account for the existence of the genetic code without resorting to the Schrödinger solution of *new laws of physics*. To this purpose, Pattee focussed on the idea that physical theory does not consist only of physical laws, but of laws *plus* initial conditions and boundary conditions, both of which are often referred to as *constraints*.

This had been known since Newton's time, of course, but physicists had consistently assumed that laws are fundamental whereas constraints have only an accessory role. The reality, however, turned out to be very different. Murray Gell-Mann (1994) underlined that the effective complexity of the universe receives only a small contribution from the fundamental laws. The rest comes from 'frozen accidents', which are precisely the result of constraints. All planets, for example, are formed according to universal physical laws, and yet they are all different. Their individual features are due to the particular constraints of their historical development, and the distinction between laws and constraints is so important that Eugene Wigner (1964) called it "Newton's greatest discovery".

In this novel theoretical framework where laws and constraints have equally fundamental roles, Pattee argued that information and codes are perfectly compatible with physical theory because they have precisely the defining features of constraints. The rules of a code, for example, are limitations that drastically reduce the number of possibilities and can be regarded therefore as true natural constraints. In a similar way, Claude Shannon underlined that information is obtained whenever uncertainty is reduced, and concluded from this that the notions of information and constraint are interchangeable (Shannon 1948).

The solution proposed by Pattee, in short, is that information and codes do not require new laws of physics, because they are *a special type* of constraints and constraints are an integral part of physical theory (Pattee 1968, 1972, 2001, 2008). This is the *'special constraint'* solution to the problem of the genetic code, a solution that is developed in three logical steps: (1) life requires self-replication (a biological principle), (2) evolution requires symbolic control of self-replication (von Neumann), and (3) physics describes information and codes as special types of constraints (Pattee).

Such a conclusion, however, is not wholly satisfactory. It is certainly true that sequences and codes have the characteristics of constraints, but not all constraints lead to life, far from it, and it is not enough to say that they must be 'special' constraints. What is it that makes them special? What is it that distinguish the special constraints of information from the special constraints of the genetic code, and what is it that distinguish both of them from the countless constraints of inanimate matter?

## 2.3 The New Observables

Howard Pattee has pointed out that biology does not need new laws of physics because physical theory is based on laws and constraints, and entities like information and coding rules can be regarded as special types of constraints. This is undoubtedly true, but it is not the whole truth. Physical theory starts with the definition of fundamental entities, or *observables* (time, space, mass etc.), and then looks for relationships between them which are referred to as laws and constraints. The basic components of physical theory, in short, are not two but three: laws, constraints, and observables.

If we assume *a priori* that life does not need new observables, we can limit ourselves to laws and constraints, but this is precisely the point that we cannot take for granted. Life is based on the copying of genes and on the coding of proteins and these processes require entities, like biological sequences and the rules of a code, that have all the defining characteristics of *new observables*. This is because the role of observables is to allow us to describe the world and we simply cannot describe living systems without sequences and codes. But what kind of entities are these new observables?

A biological sequence is a linear chain of subunits that represents *organic information*, and a biological code is a set of rules that associate an *organic meaning* to organic molecules. Sequences and codes, in short, are carriers respectively of organic information and organic meaning, and our problem is to understand the nature of these entities.

According to a long tradition, natural entities are divided into *quantities* and *qualities*. Quantities can be measured and are objective, whereas qualities are subjective and cannot be measured. In the case of organic information and organic meaning, however, this scheme breaks down. Organic information, for example, is not a quantity because the specific order of a sequence cannot be measured. But it is not a quality either, because linear specificity is a feature that we find in organic molecules, and is therefore an objective feature of the world, not a subjective one. The same is true for organic meaning. This too cannot be measured, so it is not a quantity, but it is not a quality either because the rules of the genetic code are the same for all observers in all living systems.

A scheme based on quantities and qualities alone, in short, is not enough to describe the world. In addition to quantities (*objective and measurable*) and qualities (*subjective and not-measurable)* we must recognize the existence in Nature of a third type of entities (*objective but not-measurable*).

Organic information and organic meaning belong precisely to this new type of entities, and we can also give them a suitable name. Since organic information and organic meaning can be described only by *naming* their components, we can say that they are *nominable* entities, or that they belongs to the class of the nominable entities of Nature (Barbieri 2004, 2006a, 2008).

It must be underlined that the existence of new observables in all living systems is perfectly compatible with physics, because observables are an integral part of physical theory and the discovery of new observable has gone on throughout the history of science. Let us take therefore a closer look at these new natural entities and see if we can learn something more about them.

## 2.4 Names and 'Nominable' Entities

Physical theory consists of laws, constraints and observables, but in addition to these three components there is also a fourth one that should be taken into account, and that is *names*. Science is always expressed in words and we need therefore to give names to the objects and the processes that we observe in Nature. Names (including those that we call 'numbers') are necessarily a fourth essential component of physical theory, but are different from the first three because they change from one language to another. Laws, constraints and observables, in other words, do not depend upon the language that is employed to express them, whereas names are totally language-dependent. This is because names (or *nominal entities*, to use a classical term) in general have nothing to do with the intrinsic features of the named objects, and are therefore labels that can be attached to them.

The deep divide that exists between 'names' and 'objects' has been at the centre of many controversies in the past, in particular of the celebrated medieval dispute over 'nominal entities' and 'real entities'. It has also had a long history in the philosophy of mathematics, where some have argued that numbers are 'invented' by the human mind, and others that they are 'discovered', a conclusion which implies that they have an existence of their own in some abstract Platonic world.

The relationship between names and objects is also a crucial issue in science, but here it has taken on a new form. Let us underline that all names are sequences of characters (alphabetic, numerical or alpha-numerical) and that each sequence is *unique*. Names, in other words, have *specificity*. In general, the specificity of a name has nothing to do with the characteristics of the named object, and in these cases we can truly say that names are mere labels. Science, however, has invented a new type of names where the sequence of characters does represent an order that is objectively present in the named objects.

The chemical formula of a molecule, for example, describes an objective sequence of atoms, and any atom can be described by the objective sequence of its quantum numbers. In these cases, the names are no longer arbitrary labels but true '*observables*' because they describe characteristics that we observe in Nature. This shows that there are two distinct types of names in science: labels and observables.

In the case of the observables, furthermore, there is another distinction that must be considered. When a molecule is formed spontaneously, its final sequence is due to the interactions between its own components, and in most cases it is completely determined by them. In the case of a protein, however, all its different amino acids interact by the same peptide bonds and a spontaneous assembly would produce a completely random order (which is incompatible with life). In this case, a specific sequence can be obtained only if the amino acids are put together by a molecular machine according to the order provided by a template that is *external* to the protein itself. We need therefore to distinguish between two different types of observables.

The sequence of quantum numbers in an atom, or the sequence of atoms in inorganic molecules, is determined *from within*, by internal factors, whereas the sequence of amino acids in a protein is determined *from without,* by external templates. In the first case the sequence is a *physically computable* entity, in the sense that it is the automatic result of physical forces, whereas in the second case it can only be described by 'naming' its components, and is therefore a *nominable* entity (this term should not be confused with the classical concept of *nominal* entity, which applies to all names). A *nominable* entity is not a label but an observable, and more precisely a *non-computable* observable.

All names, in conclusion, are specific sequences of characters, and in science they can be divided into two great classes: labels and observables. The observables, in turn, can be divided into *computable* entities and *nominable* entities. The important point is that physics and chemistry deal exclusively with computable entities (physical quantities), whereas nominable entities (information and coding rules) exist only in living systems. We need therefore to pay a special attention to these new observables, and make sure that they truly are fundamental entities of Nature.

## 2.5 Organic Information

In genes and proteins, biological, or organic, information has been defined as the specific sequence of their subunits. This definition however is not entirely satisfactory because it gives the impression that information is a *static* property, something that molecules have simply because they have a sequence. In reality, there are countless molecules which have a sequence but only in a few cases this becomes information. That happens only when a copymaker uses it as a guideline for copying. Even copymakers, however, do not account, by themselves, for information. Copymakers produce sequences by sticking subunits together, but without a template they would produce only *random* sequences, not specific ones. Sequences alone or copymakers alone, in other words, have nothing to do with information. It is only when a sequence provides a guideline to a copymaker that it becomes information for it. It is only an act of copying, in other words, that brings organic information into existence.

This tells us that organic information is not just the specific sequence of a molecule, but *the specific sequence produced by a copying process*. This definition underlines the fact that organic information is not a thing or a property, but the result of a process. It is, more precisely, an 'operative' definition, because information is defined by the process that brings it into existence. We realize in this way that organic information is as real as the copying process that generates it.

We have also seen that organic information is neither a quantity (because a specific order cannot be measured), nor a quality (because it is an objective feature of all copied molecules), and belongs instead to a third class of objects that have been referred to as *nominable* entities (Barbieri 2004, 2006a, 2008).

We conclude that organic information is a new type of entity and that it is essential to describe the basic molecules of life. To this purpose, in fact, it is no less essential than the physical quantities, and this means that organic information *has the same scientific status as a physical quantity*. They both belong to the class of objective and reproducible entities that allow us to describe the world.

This conclusion, however, raises immediately a new problem, because there are two distinct groups of physical quantities: a small group of *fundamental* quantities (space, time, mass, charge and temperature) and a much larger group of *derived* quantities. That distinction applies to all objective entities, so we need to find out whether organic information belongs to the first or to the second group.

Luckily, this problem has a straightforward solution because the sequences of genes and proteins have two very special characteristics. One is that *a change in a single component of a biological sequence may produce a sequence which has entirely new properties*. This means that although a biological sequence can be said to have 'components', it is at the same time a single indivisible whole. The second outstanding feature is that *from the knowledge of n elements of a biological sequence we cannot predict the element (n + 1)*. This is equivalent to saying that *a specific sequence cannot be described by anything simpler than itself*, so it cannot be a derived entity.

We conclude that organic information has the same scientific status as the physical quantities, because it is an objective and reproducible entity. But we also conclude that it does not have the status of a derived physical quantity because it cannot be expressed by anything simpler than itself. This means that organic information has the same scientific status as the fundamental quantities of physics, and is therefore a new irreducible entity of Nature, i.e., a new fundamental observable.

## 2.6 Organic Meaning

A code is a set of rules that establish a correspondence between the objects of two independent worlds. The Morse code, for example, is a correspondence between groups of dots and dashes with the letters of the alphabet, and in the same way the genetic code is a correspondence between groups of nucleotides and amino acids. Let us notice now that establishing a correspondence between, say, object 1 and object 2, is equivalent to saying that object 2 is the *meaning* of object 1. In the Morse code, for example, the rule that 'dot-dash' corresponds to the letter 'A', is equivalent to saying that letter 'A' is the meaning of 'dot-dash'. In the code of the English language, the sound 'apple' is associated to the fruit 'apple', and this is equivalent to saying that that fruit is the meaning of that sound.

By the same token, the rule of the genetic code that a group of three nucleotides (a codon) corresponds to an amino acid is equivalent to saying that that amino acid is the *organic meaning* of that codon. Anywhere there is a code, be it in the mental or in the organic world, there is meaning. We can say, therefore, that *meaning is an entity which is related to another entity by a code*, and that organic meaning exists whenever an organic code exists (Barbieri 2003; Artmann 2007, 2009).

The existence of meaning in the organic world may seem strange, at first, but in reality it is no more strange than the existence of a code because they are the two sides of the same coin. To say that a code establishes a correspondence between two entities is equivalent to saying that one entity is the meaning of the other, so we cannot have codes without meaning or meaning without codes. All we need to keep in mind is that *meaning is a mental entity when the code is between mental objects, but it is an organic entity when the code is between organic molecules.*

Modern biology has readily accepted the concept of information but has carefully avoided the concept of meaning, and yet organic information and organic meaning are both the result of natural processes. Just as it is an act of copying that creates organic information, so it is an act of coding that creates organic meaning. Copying and coding are the processes; copymakers and codemakers are their agents; organic information and organic meaning are their results.

But the parallel goes even further. We have seen that organic information cannot be measured, and the same is true for organic meaning. We have seen that organic information is an objective entity, because it is defined by the same sequence for any number of observers, and that is also true for organic meaning, which is defined by coding rules that are the same for all observers. Finally, we have seen that organic information is an irreducible entity, because it cannot be described by anything simpler than its sequence, and the same is true for organic meaning, which cannot be defined by anything simpler than its coding rules.

Organic information and organic meaning, in short, belong to the same class of entities because they have the same defining characteristics: they both are *objective-but-not-measurable* entities, they both are *fundamental* entities because they cannot be reduced to anything simpler, and they both are *nominable* entities because we can describe them only by naming their components (Barbieri 2004, 2006a, 2008).

Finally, let us underline that they are the twin pillars of life because organic information comes from the copying process that produces genes, while organic meaning comes from the coding process that generates proteins.

## 2.7 The Discovery of New Worlds

The history of physics tells us that scientific discoveries require three logical steps. First we look at the world and choose a certain number of entities to describe it, entities that are called *observables* (space, time, mass, etc.) precisely because they represent what we observe. Then we look for relationships between observables and obtain models of the observed phenomena (regularities, equations, laws, etc.). Finally we use our models to make predictions that test them (we predict, for example, the nest eclipse of the moon etc.).

The choice of the observables is the first step in the procedure and the most critical. The movements of planets and stars, for example, can be described with only two observables –space and time– and in that case we get either a Ptolemaic model or a Copernican system. By introducing a third observable – mass – we obtain the laws of motion, universal gravitation and the Newton model of the world.

The three basic observables of classical physics can be combined together in different ways and produce many other derived observables (speed, acceleration, force, energy, power, momentum, etc.), but what defines the whole system is the initial number of fundamental observables. The actual identity of these observables can be changed (space and time, for example, can be replaced by speed and time, and in that case space becomes a derived entity), but the minimum number of fundamental observables does not change. That number defines a whole world of phenomena, and we can discover new worlds, i.e., new aspects of reality, only if we discover new fundamental observables. The world of electricity and magnetism, for example, required precisely the introduction of new fundamental observables, and so did the world of thermodynamics, the world of nuclear forces, and the world of elementary particles. All of which takes us to a question: do we need new observables in the world of life or not? This point is crucial, and the different paradigms of biology are nothing less than different ways to answer it.

The chemical paradigm states *a priori* that we do not need new observables to describe living systems, i.e., that life is completely described, in principle, by the quantities of physics. The information paradigm claims that information is a *fundamental* entity that exists only in living systems, but it has not been able to contrast the physicalist charge that there is nothing fundamental in it.

We can prove that this charge is wrong only by showing that information is a new *observable* and this can be done only by showing that information is the result of a manufacturing process by molecular copying. But as soon as we accept the reality of molecular copying we must also accept the reality of molecular coding, and therefore of another fundamental observable. This is the third paradigm of modern biology, the Code view of life, the idea that life is artifact-making by copying and coding.

The crucial point is that the existence of two new observables in living systems is not a hypothesis. It is an *experimental* fact. We can prove that biological sequences (organic information) and the rules of a code (organic meaning) are fundamental observables with the same procedures that we have used in the case of space, time, mass, temperature, etc. The only difference is that sequences and coding rules are *non-computable* observables, but there is no doubt that observables they are (we do observe them in living systems) and that they are *fundamental* observables (because we cannot describe living systems without them and because we cannot reduce them to anything else).

Classical physics, thermodynamics, electromagnetism and elementary particles, were all based on the discoveries of new fundamental observables, and now we realize that this is also true in biology. Life is indeed a new world, a new dimension of reality, because it is the result of copying and coding processes that bring new fundamental observables into existence.

We conclude in this way that life is distinct from chemistry because to describe the world of life we need new fundamental observables *in addition* to those that describe the world of chemistry.

## 2.8 Life and Semiosis

Culture is made of conventional rules, or codes, and one of its most distinctive features is the production of artifacts. Molecular biology has proved that there is a genetic code in every cell, and that genes and proteins are molecular artifacts because they are manufactured by molecular machines. Coding and artifact-making, in other words, take place both in our society and inside the cell, and this does create a parallel between culture and molecular biology. But how deep it is? What are the similarities and the differences between molecular codes and cultural codes? Between molecular artifacts and human artifacts?

In order to deal with this problem, let us start from the fact that culture is based on *signs* or, more precisely, on *semiosis* (the production of signs). This allows us to rephrase the question: what are the similarities and the differences between molecular semiosis and cultural semiosis? But first of all, does molecular semiosis exist? Can we truly say that semiosis takes place at the molecular level?

The answer depends, of course, on our definition of semiosis, and this unfortunately is not an easy issue. The definitions of sign and semiosis have been debated for more than a century by the two main schools of cultural semiosis (founded respectively by Ferdinand de Saussure and by Charles Peirce), but an agreement has not yet been reached. In such a situation, the most sensible thing to do is to adopt a step-by-step approach: let us start from a common sense definition of semiosis, see how far we can go with it, and be prepared to generalize it when we find that it is no longer effective. This suggests that there may be different types of semiosis at different levels, an idea which is worth considering because it has at least two advantages. The first is that it makes sense from an evolutionary point of view, because we can hardly expect that the first forms of semiosis which appeared on Earth had the full blown complexity of human semiosis. The second advantage is that it is in line with the solution that biologists have adopted in other cases, for example in the definition of *species*.

Tradition and popular wisdom tell us that a species is any group of organisms that can produce fertile offspring, and this common sense definition, formulated in proper academic terms, has become known as *'the biological species concept'*. It is routinely used in many biological fields, but it is not universally valid. It does not help, for example, with *extinct* species, so it cannot be used to build phylogenetic trees. For that purpose biologists had to resort to other definitions, in particular to cladistic criteria, a procedure which has allowed them to overcome the terminological impasse and make real progress.

A similar procedure can be adopted also in the case of semiosis and here we will follow precisely that approach. Definitions, after all, are tools, not ends-in-themselves, and should be used as practical guidelines, not as excuses for endless terminological disputes. We shall start therefore from a *minimal* definition that can be regarded as a sort of *'biological semiosis concept'* and show that organic semiosis does exist in the cell.

## 2.9 The Code Model of Semiosis

Semiosis is *the production of signs*, and semiotics is usually referred to as *the study of signs* (from the Greek *semeion* = sign) but these definitions are too restrictive because signs are always associated with other entities. A sign, to start with, is always linked to a *meaning*. As living beings, we have a built-in drive to make sense of the world, to give meanings to things, and when we give a meaning to something, that something becomes a sign for us. Sign and meaning, in other words, cannot be taken apart because they are the two sides of the same coin. Semiotics, therefore, is not just the study of signs; it is the study of signs and meanings together. The result is that a system of signs, i.e., a *semiotic system*, is always made of at least two distinct worlds: a world of entities that we call *signs* and a world of entities that represent their *meanings*.

The link between sign and meaning, in turn, calls attention to a third entity, i.e., to their *relationship*. A sign is a sign only when it stands for something that is *other than itself*, and this *otherness* implies at least some degree of *independence*. It means that there is no deterministic relationship between sign and meaning. Different languages, for example, give different names to the same object precisely because there is no necessary connection between names and objects. A semiotic system, therefore, is not any combination of two distinct worlds. It is *a combination of two worlds between which there is no necessary link*, and this has an extraordinary consequence. It implies that a bridge between the two worlds can be established only by *conventional* rules, i.e., by the rules of a *code*. This is what qualifies the semiotic systems, what makes them different from everything else: *a semiotic system is a system made of two independent worlds that are connected by the conventional rules of a code*. A semiotic system, in conclusion, is necessarily made of at least *three* distinct entities: *signs, meanings* and *code*.

Here at last we have a definition where it is stated explicitly that a code is an essential component of a semiotic system. It is the rules of a code that create a correspondence between signs and meanings, and we can say therefore that an act of semiosis is always a conventional act. More precisely, we can say that an *elementary* act of semiosis is a triad of *'sign, meaning and convention'*, whereas a semiotic system is a whole set of signs and meanings that are linked together by all the conventional rules that make up a code.

Signs, meanings and conventions, however, do not come into existence of their own. There is always an 'agent' that produces them, and that agent can be referred to as a *codemaker* because it is always the making of a code that gives origin to semiosis. In the case of culture, for example, the codemaker is the human brain, since it is the brain that produces the mental objects that we call signs and meanings and the conventions that link them together. We come in this way to a general conclusion that can be referred to as 'the Code model of semiosis': *a semiotic system is a set of signs, meanings and code that are all produced by the same agent, i.e., by the same codemaker*.

This conclusion is highly relevant to biology because it tells us precisely what we need to prove in order to show that the cell is a semiotic system. We need to prove that in every living cell there are four distinct entities: signs, meanings, code and codemaker.

## 2.10 The Defining Feature of Signs and Meanings

A semiotic system is made of signs, meanings, code and codemaker, and we know that there is a genetic code in protein synthesis. We also know that there is a *codemaker* in the cell, because proteins are made by an internal system of ribonucleoproteins that is the physical seat of the genetic code This tells us that every living cell does have a genetic code and a codemaker. But what about the other two entities? Can we say that there are also signs and meanings at the molecular level? Can these entities exist in the cell? In order to answer this question, let us examine first the traditional signs and meanings of culture and see if they have a qualifying feature that can be extended to the molecular level.

The signs and meanings that we are familiar with are often related to objects or events of the physical world. A sign, for example, can be a spoken word and its meaning can be a mental reaction. The mental (or neural) image of an object is normally evoked by different words in different languages, and this clearly shows that mental sounds and mental images are separable. When they are separated, however, they no longer function as signs and meanings. To a non-English speaker, for example, a word like 'twitch' may have no linguistic meaning and in this case it would be just a sound, not a sign. There is no contradiction therefore in saying that signs and meaning are distinct mental objects and that they cannot be taken apart, because when they are taken apart they simply stop functioning as signs and meanings.

This makes us understand an extremely important feature of semiosis. It tells us that a mental sign, or a mental meaning, is never an *intrinsic* property of a mental object. It is something that the mind can give to a mental object and that the mind can take away from it. This conclusion remains valid even if we accept that the mind is but a product of the brain. In this case, in fact, the link between the neuron firings that produce the signs and the meanings of any language are based on a neural code and are totally dependent on the 'agent' of that code, i.e., on the neural *codemaker* of the system. We realize in this way that *signs and meanings simply do not exist without a codemaker and outside a codemaking process*. The codemaker is the *agent* of semiosis, whereas signs and meanings are its instruments.

We conclude that signs and meanings are totally dependent on codemaking, i.e., that they are *codemaker-dependent entities*. This is the qualifying feature that we were looking for, because it is completely general and can be applied to all systems. We can say therefore that signs and meanings exist at the molecular level, and in particular in protein synthesis, only if we prove that in protein synthesis there are codemaker-dependent entities.

## 2.11 The Sequences of Genes and Proteins

All biochemistry textbooks tell us that there is a genetic code in protein synthesis, but none of them mentions the existence of signs and meanings. At first sight, in fact, these entities do not seem to exist at the molecular level. The translation apparatus can be regarded as a codemaker because it is the seat of the code that creates a correspondence between genes and proteins, but these molecules appear to have only 'objective' chemical properties, not the 'codemaker-dependent' properties that *define* signs and meanings. A messenger RNA, for example, appears to be a unique and objective sequence of molecules, but let us take a closer look.

A messenger RNA is certainly a unique and objective chain of *nucleotides* but in no way it is a unique sequence of *codons* because different codemakers could scan it in different ways. If the nucleotides were scanned two-by-two, for example, the sequence of codons would be totally different. The same chain of nucleotides, in other words, can give origin to many sequences of codons, and it is always the codemaker that determines the sequence because it is the codemaker that *defines* the codons. A linear sequence of codons, in short, does not exist without a codemaker and outside a codemaking process. It is totally dependent on codemaking and is therefore a *codemaker-dependent entity*, which is precisely what a sign is.

In the same way, the linear sequence of amino acids that is produced by the translation apparatus is also a codemaker-dependent entity, because only a codemaker can produce it. Any spontaneous assembly of amino acids would *not* make linear chains, and above all it would not arrange the amino acids in a specific order. Specific linear sequences of amino acids can be produced only by codemakers, but different codemakers would arrange the amino acids in different ways, which shows that the sequence of a protein is only one of the many possible 'meanings' that could be given to a string of nucleotides.

The sequence of a gene and the sequence of a protein, in conclusion, are not objective properties of those molecules. They are codemaker-dependent properties because they do not exist without a codemaking process, and because they would be different if the codemaker had different properties. The sequences of genes and proteins, in short, have precisely the characteristics that define signs and meanings. They are codemaker-dependent entities made of organic molecules and are therefore organic signs and organic meanings. All we need to keep in mind is that *signs and meanings are mental entities when the codemaker is the mind, but they are organic entities when the codemaker is an organic system* (Barbieri 2003).

We reach in this way the conclusion that every living cell contains all four components of semiosis (signs, meanings, code and codemaker) and is therefore a real semiotic system.

## 2.12 Two Types of Organic Signs

Signs have been divided since antiquity into two great classes that are traditionally represented by *symbols* and *symptoms*. Augustine (389ad) called them *signa data* and *signa naturalia*, a distinction that continues to these days under the terms of *conventional signs* and *natural signs* (Favareau 2007).

The conventional signs are those where no *physical* relationship exists between signifiers and signified, and a connection between them can be established only by arbitrary rules, i.e., by conventions. Words, for example, are signs (because they 'stand for' the named entities) and are conventional signs because they are not determined by the characteristics of the named entities. In the same way, there is no necessary connection between symbols and the entities that they stand for (between a flag and a country, for example).

In natural signs, by contrast, a physical link is always present between signifier and signified, but it can be of different types and for this reason the natural signs have been divided into two major classes called *icons* and *indexes*. Icons are signs where a similarity exists between signifiers and signified (a portrait or a caricature, for example, is an iconic sign of a person). Indexes are signs that point to, or 'indicate', a signified object (smoke as a sign of fire, odours as signs of food, footprints as signs of organisms, etc.).

In all natural signs (icons and indexes) there is always some physical relationship between the visible signs and the invisible entities that they stand for, and yet the relationship is *underdetermined*, so much so that it takes a process of learning and an act of interpretation to establish it. The diagnosis of an illness from symptoms, for example, is always an interpretive exercise, and even simple associations, such as those between clouds and rain, depend upon learning and memory.

At the molecular level we have seen that the apparatus of protein synthesis uses a sequence of nucleotides to produce a sequence of amino acids according to the rules of the genetic code. In that case, there is no necessary connections between nucleotides and amino acids and the sequence of nucleotides is used therefore as a *conventional* organic sign, i.e., as an *organic symbol.*

A sequence of nucleotides, however, can also be used by a polymerase to produce a *complementary copy* of that sequence, and in that case the relationship between the two molecules is no longer established by a code but by direct physical interactions between complementary surfaces. Such interactions, however, occur between very small regions of the molecules, and this means that the first sequence provides only a limited number of physical determinants for the second. The first sequence, in other words, does have a physical relationship with the second, but such relationship is underdetermined and represents only an *organic icon* for the second.

We realize in this way that natural and conventional signs also exists at the molecular level and are involved in different molecular processes. Sequences of nucleotides are used as natural signs in molecular *copying*, and as conventional signs in molecular *coding*. The copying of genes and the coding of proteins, in other words, are two distinct types of semiosis, one based on iconic signs (copying semiosis) and one based on symbolic signs (coding semiosis).

The origin of molecular copying and the origin of molecular coding produced the first two major transitions in the history of life, the transitions that gave origin to genes and proteins, but they also produced the first two types of semiosis, and this makes us realize that the origin of life was also the origin of organic semiosis, i.e., that life and semiosis are *coextensive*.

## 2.13 A New Beginning

The idea that life is an extremely complex form of chemistry is still very popular, today, and is based on the physicalist thesis that all biological processes can be reduced, in principle, to physical quantities. According to this view, genetic information and the genetic code are not fundamental entities because they are not physical quantities. They are regarded instead as metaphorical and teleological terms that we use only because they are intuitively appealing.

We have seen however that the physicalist thesis is valid only in spontaneous systems, whereas genes and proteins are never formed spontaneously in real life. They are invariably manufactured by molecular machines, and all manufacturing processes do not require only physical quantities but also additional entities like sequences and coding rules.

The charge that information is a teleological concept is simply false, notwithstanding the fact that it is repeated fairly often. The truth is precisely the other way round. Information has all the defining features of a scientific concept because it has been defined in two different ways and in both cases there is nothing teleological about it.

1. When it is defined by Shannon's approach, information is actually expressed by a formula, like any standard physical quantity.
2. When it is defined by a sequence, information is no longer measurable, but it still is a fundamental observable because it is absolutely necessary to the *description* of living systems.

We simply cannot describe the transmission of genes or the synthesis of proteins without their sequences, and we cannot replace these sequences with anything else, which means that using information to describe living systems is perfectly equivalent to using space, time, mass and energy to describe physical systems. The truth, in other words, is that there is no more teleology in information and in the genetic code than there is in the quantities of physics and chemistry. Sequences (biological information) and coding rules (biological meaning) are *descriptive* entities and are absolutely essential to the scientific study of life.

The information paradigm, on the other hand, has claimed that information is distinct from chemistry but has not been able to say why. On top of that, it has accepted the concept of information but not the concept of meaning, which is equivalent to saying that genetic information is real but the genetic code is not, again without being able to say why.

We need therefore a new paradigm that fully accepts the implications of the discovery of the genetic code. The implication that life is based on copying *and* coding. This is the code paradigm, the theoretical framework where biological sequences (organic information) and biological coding rules (organic meaning) are *fundamental* observables that are as essential to life as the fundamental quantities of physics.

These theoretical concepts are the precondition for a new paradigm, but they are not enough. We also need the experimental proof that the genetic code was not an extraordinary exception but the first of a long series of organic codes. We need the demonstration that many other organic codes exist in Nature and are an integral part of the fabric of life.

# Chapter 3
# A Gallery of Organic Codes

The discovery of the genetic code has revealed the existence of a deep parallel between protein synthesis and language. In both cases, a small set of units is used to create an unlimited variety of objects by arranging the units in countless different combinations. In technical terms this process is called *recursion*, and the presence of recursion at the molecular level raised the possibility that there is a *molecular language* at the basis of life (Beadle and Beadle 1966). At the same time, however, the first models of the genetic code were all based on the *stereochemical hypothesis*, the idea that the coding rules are dictated by chemical relationships in three dimensions, whereas language is made of *arbitrary* conventions. Eventually, however, the stereochemical hypothesis had to be abandoned because it became clear that the rules of the genetic code are not the result of chemical necessity. In this sense they are as arbitrary as the rules of language, and this makes us realize that at the molecular level there is not only recursion but also *arbitrariness*.

It turned out, furthermore, that the genetic code is not an isolated case. Various other organic codes have been discovered in living systems, and the stereochemical hypothesis had to be ruled out in all of them. The evidence, in other words, tells us that codes consist of arbitrary rules not only in the cultural world but also at the molecular level. There are various important differences between organic codes and cultural codes, but arbitrariness is not one of them. On the contrary, it is the property that best reveals the existence of coding rules and represents a defining feature in all codes.

This allows us to say that a code is *a set of rules that establish a mapping between two independent worlds*, where two worlds are said to be independent when there is no necessary connection between them and a mapping can be established in countless different ways by arbitrary rules. This in turn means that a *specific* mapping between the two worlds can only be achieved by selecting a fixed number of coding rules.

At the molecular level, a mapping between two independent types of molecules is necessarily realized by a third type of molecules that act as intermediaries, and it is these molecules, called *adaptors*, that reveal the presence of a code. The existence

© Springer International Publishing Switzerland 2015

M. Barbieri, *Code Biology*, DOI 10.1007/978-3-319-14535-8_3

of organic codes in Nature, in short, is quintessentially an experimental problem, and the purpose of this chapter is to illustrate this point in the case of some of the major organic codes that have been discovered so far.

## 3.1 The Apparatus of Protein Synthesis

In 1944, Avery, Macleod and McCarty announced that the transformation of pneumonia bacteria from a mild form into a virulent one (a phenomenon discovered by Fred Griffith in 1928) is caused by almost pure DNA, a result which implied that DNA is the sole responsible of hereditary transformations. This took biologists by surprise because it was widely assumed that nucleic acids have boring and repetitive structures, much like starch and cellulose, that prevent them from carrying information. Avery was aware that his results required a new theory of heredity, and said so explicitly at the end of the paper: "*If the results of the present study on the nature of the transforming principle are confirmed, then nucleic acids must be regarded as possessing biological specificity the chemical basis of which is as yet undetermined*" (Avery et al. 1944).

Avery's paper was generally received with skepticism, but then came another set of results that pointed in the same direction. In 1946, Jean Brachet argued that a large number of experimental data strongly suggests that "*DNA is primarily confined to the nucleus, while RNA is mainly found in the cytoplasm, and protein synthesis is associated with RNA*". More precisely, Brachet (1944, 1946) argued that protein synthesis takes place on heavy cytoplasmic particles that he called *ribonucleoprotein granules*.

Shortly afterwards, André Boivin and Roger Vendrely (1947) carried the scheme of Brachet to his logical conclusion and proposed that '*DNA makes RNA makes Proteins*'. This 1947 version of the Central Dogma was, like Avery's discovery, largely ignored, but the idea was born and the evidence started accumulating.

In 1952, Alexander Dounce made another revolutionary proposal. He suggested that there must be a *code of correspondence* between nucleic acids and proteins, and since there are 20 amino acids but only 4 nucleotides, he proposed that each amino acid is coded by a group of three nucleotides (a *codon*). On top of that, Dounce (1952, 1953) proposed that the attachment of the 20 canonical amino acids to nucleotides is promoted by 20 specific catalysts that he called *activating enzymes*. After their discovery, these have been named *aminoacyl-tRNA-synthetases*, or, more briefly, *synthetases*.

Between 1944 and 1952, in short, the basic concepts of molecular biology had already been formulated by Avery, Brachet and Dounce, but apparently did not raise much interest. The turning point came in 1953, when James Watson and Francis Crick proposed the double helix, the model that in one swift stroke illuminated the structure of DNA and revealed that hereditary information is carried by linear sequences of nucleotides (Watson and Crick 1953).

In 1957 Francis Crick argued there must be intermediary molecules between nucleotides and amino acids, molecules that are made of RNA and that recognize a codon by a complementary sequence of nucleotides that he called *anticodon*. Crick (1957) called these molecules *adaptors*, but in that same year they were discovered by Hoagland et al. (1957) and became known as *transfer-RNAs*.

A year later, Francis Crick (1958) re-proposed the Central Dogma of molecular biology, and this time the idea was immediately accepted: *DNA makes RNA* (transcription) *and RNA makes proteins* (translation). In that same year, Roberts (1958) gave the name *ribosomes* to the supramolecular machines of protein synthesis (the ribonucleoprotein granules described by Brachet), and the conclusion that *the ribosome is the decoder of genetic information* acquired the status of an experimental fact.

Ribosomes account for more than 80 % of the total RNA of a cell, and it was taken for granted at the time that the information of the genes is transported by ribosomal-RNAs, but things turned out very differently. In 1961, François Jacob and Jacques Monod proved that the carriers of genetic information are a completely different family of molecules that they called *messenger-RNAs* (Jacob and Monod 1961). Later on, it was discovered that the scanning of the messenger-RNAs requires a whole new battery of enzymes that were called *initiation*, *elongation* and *termination factors* (Nomura et al. 1974).

The apparatus of protein synthesis, in conclusion, consists of ribosomes, messenger-RNAs, transfer-RNAs, synthetases, amino acids and scanning factors. It is a huge supramolecular system made of more than 120 types of components, and it is by far the most complex apparatus that exists at the cellular level.

## 3.2 The Genetic Code

A messenger RNA is scanned by a ribosome in groups of three nucleotides, called *codons*, and every codon is recognized by the *anticodon* of a transfer-RNA that carries one of the 20 canonical amino acids. The total number of codons that can be obtained with 4 nucleotides (A, U, C, G) is 64 ($4^3$), and the rules of correspondence between 64 codons and 20 amino acids represent, collectively, *the genetic code*.

In 1961, Niremberg and Matthaei announced that an artificial messenger containing only uracil (U) is translated into a protein that contains only phenylalanine (Phe), which means that the codon UUU codes for phenylalanine (Niremberg and Matthaei 1961). They had deciphered the first codeword of the genetic code. Other codewords were identified with artificial messengers made of nucleotides arranged in various orders (Speyer et al. 1963). It was found, for example, that a sequence of alternating uracil and cytosine (. . . UCUCUCUC . . . ) codes for a polypeptide made of alternating serine and leucine ( . . . Ser-Leu-Ser-Leu . . . ), thus proving that UCU is a codon for serine and CUC is a codon for leucine (Niremberg and Leder 1964; Nishimura et al. 1965). Various other techniques were designed to unlock the meaning of the other codons and by 1966 the genetic code was completely deciphered (Khorana et al. 1966; Niremberg et al. 1966).

It turned out that 61 codons code for amino acids, and that one of them (usually AUG) is also used as a start signal, whereas the remaining three (UAA, UAG and UGA) are termination signals. Between 61 codons and 20 amino acids there is necessarily a many-to-one correspondence, and this is expressed by saying that the genetic code is *degenerate* (or *redundant*). More precisely, some amino acids are specified by 6 codons, some by 4, others by 2, and only 2 amino acids are coded by a single codon. The genetic code is therefore redundant but *not ambiguous* because anyone of the 61 codons codes for one and only one amino acid.

In principle, this implies that every cell contains 61 different types of tRNAs, one for each codon, but in practice the actual number is about 40 per cell. The best explanation for this surprising fact was proposed by Crick (1966) with what has become known as the *wobble hypothesis*. Crick pointed out that the three nucleotides of an anticodon stick out like fingers from the surface of their t-RNA and this allows them to oscillate, or *wobble*. The result is that a nucleotide in an anticodon (especially in third position) can form a temporary bond in the codon not only with its complementary nucleotide but also with another one with which it has a partial similarity. A uracyl in third position, for example, can form a bond not only with adenine but also with guanine, and in this case its tRNA can associate the same amino acid to two distinct codons. This means that only one tRNA, rather than two, is sufficient when two codons specify the same amino acid and this is indeed what happens in various cases. It has been found, for example, that the codons that end with adenine (XYA) and those that end with guanine (XYG) code for the same amino acid, and the same is true for the codons that end with uracyl (XYU) and with cytosine (XYC). Lisine, for example, is codified by AAA and AAG, whereas tyrosine is codified by UCU and UAC.

There are, in short, some regularities in the genetic code that allow a cell to carry far less than 61 different types of tRNAs, and this suggests that there has been an evolution of the code towards an optimization of its performance. Such a conclusion has also been obtained with other arguments. Carl Woese, for example, proposed that the code has been optimized for minimizing the impact of translation errors (Woese 1965; Woese et al. 1966), and various statistical studies have shown that the genetic code has been optimized in respect to countless potential alternatives (Haig and Hurst 1991; Freeland and Hurst 1998; Bollenbach et al. 2007).

By far the most outstanding feature of the genetic code has come from the discovery that its coding rules are virtually the same in all living creatures. A few exceptions do exist (in the yeast *Candida*, for example, CUG codes for serine rather than leucine) but they are very minor changes and occur in an infinitesimal number of organisms. The genetic code, in other words, is virtually universal, and this means that has it been transmitted to all living systems by the primitive population that first evolved it, a population that has become known as the *common ancestor*.

## 3.3 Stereochemistry and Arbitrariness

In 1954, George Gamow proposed the *stereochemical hypothesis*, the idea that the genetic code is the result of direct interactions in space between nucleotides and amino acids. More precisely, he proposed that the amino acids fit with a lock-and-key mechanism into 'holes' formed by four nucleotides, a reaction that he called *stereochemical* because he assumed that it is the three-dimensional shape of each hole that determines which amino acid can bind to it. He added that two of the four nucleotides in every hole are complementary, which implies that each amino acid is effectively identified by a triplet.

This *diamond code* proposed by Gamow (1954) was soon found in disagreement with the evidence, but the stereochemical hypothesis was reproposed in many other ways, even when it became clear that there are molecular intermediaries between codons and amino acids (Pelc and Weldon 1966; Melcher 1974; Shimizu 1982). Today it is known that codons are recognized by anticodons carried by transfer-RNAs, and that the transfer-RNAs receive amino acids from molecules called *aminoacyl-tRNA-synthetases*.

There are 20 different synthetases, one per amino acid, and each synthetase donates its amino acid to one or more transfer-RNAs that are recognized by a small number of loci, or *identity marks*, that are distributed in various places on the surface of the tRNAs. The crucial point is that the recognition of the amino acids is independent from the anticodons because it has been shown that in many cases the synthetases have no access at all to the anticodons (Schimmel 1987; Schimmel et al. 1993).

The synthetases, in other words, are true *adaptors* because they perform two *independent* recognition processes, and establish connections that could have been realized in many other different ways. This has been proved by a variety of experiments. Hou and Schimmel (1988), for example, managed to introduce two extra nucleotides in a tRNA without changing its anticodon, and found that the resulting tRNA was recognized by a different synthetase and was carrying therefore a different amino acid. They had changed one of the rules of the genetic code, a result that achieved *in vitro* what a few microorganisms have achieved *in vivo* in the course of evolution (Jukes and Osawa 1990, 1993).

The lesson that comes from these experiments is that the rules of the genetic code can be changed in countless different ways, because they are the result of interactions between synthetases and tRNAs that can be modified virtually at will by adding or subtracting a few molecules. This means that the number of adaptors between codons and amino acids is potentially unlimited, and only the selection of a fixed number of them can ensure a specific link. It also means that the rules of the genetic code are not dictated by stereochemistry and in this sense they are *arbitrary*.

It is worth mentioning, at this point, a fairly widespread argument according to which the rules of the genetic code cannot be arbitrary because they have been optimized in the course of evolution. In reality, the two things are not incompatible. The rules of the Morse code, for example, have been optimized by associating the most frequent letters of the alphabet with the shortest combinations of dots and dashes, and yet they continue to be arbitrary rules. An optimization of the coding rules, in other words, simply means that they are not *random*, and is perfectly compatible with their arbitrariness.

We reach in this way the conclusion, first expressed by Jacques Monod (1970), that the genetic code is *chemically arbitrary* because it is based on rules that are not dictated by physical necessity.

## 3.4 The Splicing Codes

The primary transcripts of the genes are often transformed into messenger RNAs by removing some RNA pieces (called *introns*) and by joining together the remaining pieces (the *exons*). This cutting-and-sealing operation, known as *splicing*, is a true assembly because exons are assembled into messengers, and we need therefore to find out if it is a catalyzed assembly (like transcription) or a codified assembly (like translation). In the first case splicing would require only catalysts (comparable to RNA-polymerases), whereas in the second case it would need an assembly machine and a set of adaptors (comparable to ribosome and tRNAs).

These parallels immediately suggest that splicing is a codified process because it is implemented by structures that are very much comparable to those of protein synthesis. The splicing bodies, known as *spliceosomes*, are huge molecular machines like ribosomes, and employ small molecules, known as *small-nuclear-RNAs* (*snRNAs*) which are comparable to tRNAs. The similarity, however, goes much deeper than that because splicing is carried out by molecular structures that are true adaptors. They perform two independent recognition processes, one for the beginning and one for the end of each exon, thus creating a specific correspondence between primary transcripts and messenger RNAs. Splicing, in other words, is a codified process based on adaptors and takes place with sets of rules that have been referred to as *splicing codes* (Barbieri 2003; Fu 2004; Matlin et al. 2005; Wang and Burge 2008).

It must be underlined, however, that there are two outstanding complications in splicing. One is the fact that the order in which the exons are joined together can be shuffled in various ways, an operation, called *alternative splicing*, that allows many species to generate a whole family of variant proteins from the same gene. The expression of these proteins, furthermore, can change from one tissue to another and in different stages of embryonic development, thus enormously increasing the protein variety that can be associated to a gene. Alternative splicing has in this way a powerful role in the generation of biological complexity, and splicing mistakes often have pathological effects; it has been estimated that they account for about one fifth of all inherited diseases (Buratti et al. 2006; Wang and Cooper 2007; Solis et al. 2008; Cooper et al. 2009; Tazi et al. 2009).

The other great complication of splicing is the fact that many introns carry sequences that are similar to exons but translate into nonsense and for this reason are called *pseudo exons* or *pseudo genes*. They would create havoc if incorporated into mRNAs and the splicing machinery had to evolve the means to differentiate real exons from pseudo ones. The result is that real exons contain internal identity marks that are known as *exonic splicing enhancers* (ESEs) and *exonic splicing silencers* (ESSs) (Fu 2004; Matlin et al. 2005; Pertea et al. 2007). The presence of these marks, in turn, means that the adaptors of the splicing codes are not single molecules but combinations of molecules because they must be able to recognize not only the beginning and the end of the real exons, but also their internal identity marks.

The actual deciphering of the splicing codes has already started but it is taking considerably longer than that of the genetic code because it is incredibly more complex. Let us keep in mind that the discovery of the genetic code has been facilitated by two particularly favourable features. More precisely, by the fact that (1) the adaptors are single molecules (the tRNAs) and (2) the coding units form a closed set (64 codons and 20 amino acids). In the case of splicing, instead, the adaptors are combinations of molecules (*combinatorial codes*), and the domain (or *alphabet*) of the codes is open and potentially unlimited.

The overall complexity of splicing is such that the most practical way of discovering its codes is by building computational models that are capable of predicting new splicing rules on the basis of existing data. Such models have already started appearing in the literature (Pertea et al. 2007; Barash et al. 2010; Dhir et al. 2010), and represent our first glimpse of the rules of the splicing codes.

## 3.5 The Metabolic Code

This is the first organic code that came to light after the discovery of the genetic code. It was described in *Science*, in 1975, by Gordon Tomkins, a professor of biochemistry at the University of San Francisco. Tragically, Tomkins died that very year, aged 49, from a brain tumour, and apparently his idea died with him. Recently, however, there has been an attempt to rescue his work from oblivion (Swan and Goldberg 2010) and here we will try to show that such attempt is amply justified.

Tomkins investigated the evolution of metabolism and started from the need of the ancestral cells to obtain energy. "Since both nucleic acid and protein synthesis are endergonic reactions, primordial cells were almost certainly endowed with the capacity to capture the necessary energy from the environment and to transform it into usable form, presumably ATP (*adenosine triphosphate*). The biosynthetic capabilities of primitive cells were, however, probably quite limited ... survival would therefore have required the evolution of regulatory mechanisms that could maintain a relatively constant intracellular environment in the face of changes in external conditions" (Tomkins 1975).

Granted this basic need of the cells to evolve regulatory mechanisms, Tomkins distinguished between two types of regulation that he called *simple* and *complex*, both present in modern organisms. Simple regulation is when there is a feedback relationship (positive or negative) between the components of a metabolic circuit, and the end products affecting their own metabolism.

Complex regulation is characterized by two new entities that Tomkins called *symbols* and *domains*. In order to illustrate them, Tomkins made the example of molecules that are accumulated inside a cell as a consequence of a particular environment and become a *symbol* of that environment. In most microorganisms, for example, cyclic AMP is accumulated as a result of carbon starvation and becomes a symbol of that deficiency. Another example is ppGpp (guanosine5′-diphosphate3′-diphosphate) that accumulates as a result of amino acid starvation and represents a symbol of that condition. These molecules are symbols because they bear no structural relationship to the molecules that promote their accumulation (cyclic AMP, for example is accumulated as a result of glucose starvation, but it is not a chemical analog of glucose). This is what suggested to Tomkins the existence of a *metabolic code*.

"Since a particular environmental condition is correlated with a corresponding intracellular symbol, the relationship between the extra- and intracellular events may be considered as a metabolic code in which a specific symbol represents a unique state of the environment."

Tomkins went on to show how metabolic coding in unicellular organisms might have evolved into the endocrine system of the metazoa, and described what happens in the slime mold *Dictyostelium discoideum*. "Given sufficient nutrients, this organism exists as independent myxamoebas. Upon starvation, they generate cyclic AMP and release it into the surrounding medium. This substance serves as a chemical attractant that causes the aggregation of a large number of myxamoebas to form a multicellular slug. In this case, as in *E. coli*, cyclic AMP acts as an intracellular symbol of carbon-source starvation. In addition, however, the cyclic nucleotide is released from the *Dictyostelium* cells in which it is formed and diffuses to other nearby cells, promoting the aggregation response. Cyclic AMP thus acts in these organisms both as an intracellular symbol of starvation and as a hormone which carries this metabolic information from one cell to another."

Hormones, according to Tomkins, evolved in order "to carry information from *sensor* cells in direct contact with the environment, to more sequestered *responder* cells. Specifically, the metabolic state of a sensor cell, represented by the levels of its intracellular symbols, is *encoded* by the synthesis and secretion of corresponding levels of hormones. When hormones reach the responder cells, the metabolic message is *decoded* into corresponding primary intracellular symbols. In this way, endocrine cells act as both sensors and responders, that is, intermediates in the transmission of metabolic information from primary sensor cells to the tissues in which the final chemical responses take place."

## 3.6 The Signal Transduction Codes

Living cells react to many physical and chemical stimuli from the environment, and in general their reactions consist in the expression of specific genes. We need therefore to understand how the environment interacts with the genes, and the turning point, in this field, came from the discovery that the external signals (known as *first messengers*) never reach the genes. They are invariably transformed into a different world of internal signals (called *second messengers*) and only these, or their derivatives, reach the genes. In most cases, the molecules of the external signals do not even enter the cell and are captured by specific receptors of the cell membrane, but even those that do enter (some hormones) must interact with intracellular receptors in order to influence the genes (Sutherland 1972).

The transfer of information from environment to genes takes place therefore in two distinct steps: one from first to second messengers, called *signal transduction*, and a second path from second messengers to genes which is known as *signal integration*. The surprising thing about signal transduction is that there are literally hundreds of first messengers (ions, nutrients, hormones, growth factors, neurotransmitters, etc.) whereas the second messengers belong to only four molecular families: cyclic AMP or GMP, calcium ions ($Ca^{2+}$), inositol trisphosphate (IP3), and diacylglycerol (DAG) (Alberts et al. 2007).

First and second messengers, in other words, belong to two very different worlds, and this suggests immediately that signal transduction may be based on organic codes. This is reinforced by the discovery that there is no necessary connection between first and second messengers, because it has been proven that the same first messengers can activate different types of second messengers, and that different first messengers can act on the same type of second messengers (Alberts et al. 2007). The only plausible explanation is that signal transduction is based on organic codes, but of course one would like a direct proof.

The signature of an organic code, as we have seen, is the presence of adaptors and the transmembrane receptor proteins of signal transduction do have the defining characteristics of the adaptors. The transduction system consists of at least three types of molecules: a *receptor* for the first messengers, an *amplifier* for the second messengers and a *mediator* in between (Berridge 1985). This transmembrane system performs two independent recognition processes, one for the first and the other for the second messenger, and the two steps are connected by the bridge of the mediator. This connection, on the other hand, could be implemented in countless different ways since any first messenger can be coupled with any second messenger, and this makes it imperative to have a selection in order to guarantee biological specificity.

In signal transduction, in short, we find the three defining features of a code: (1) two independents worlds of objects (first messengers and second messengers), (2) a potentially unlimited number of arbitrary connections produced by adaptors, and (3) a set of coding rules (a selection of the adaptors) that ensures the specificity of the correspondence. The effects that external signals have on cells, in short, do not depend on the energy or the information that they carry, but on the *meaning* that cells give them with sets of rules that have been referred to as *signal transduction codes* (Barbieri 1998, 2003).

One may wonder at this point why signal transduction codes are never mentioned in biochemistry books despite the fact that the their molecules are true adaptors. The problem here is that the study of signal transduction started when organic codes were not known, and it has always been assumed *a priori* that in this process there is no need for them. A code, in short, has not been found simply because it has never been looked for. The genetic code, on the contrary, was predicted on theoretical grounds, and it was discovered precisely because experiments were devised with the specific purpose to look for it.

## 3.7 The Signal Integration Codes

We have seen that there are only four families of second messengers in the cell, and yet the reactions that they set in motion can pick up an individual gene among tens of thousands. How this is achieved is still a mystery, but some progress has been made. Perhaps the most illuminating discovery, so far, is that second messengers do not act independently. Calcium ions and cyclic-AMPs, for example, have effects that in some occasions reinforce each other whereas in others are mutually exclusive (Alberts et al. 2007). The cell, in short, can combine its internal signals in countless different ways, and it is precisely this combinatorial ability that explains why a small number of second messengers can generate an extraordinarily high number of specific genetic responses. The activation of second messengers, in other words, sets in motion a cascade of reactions that normally ends with the expression of a target gene, and again we need to understand if they are normal catalized reactions or if at least some of them are based on the rules of a code.

One of the most interesting clues, in this field, is the fact that signalling molecules have in general more than one function. Epidermal growth factor, for example, stimulates the proliferation of fibroblasts and keratinocytes, but it has an anti-proliferative effect on hair follicle cells, whereas in the intestine it is a suppressor of gastric acid secretion. Other findings have proved that *all* growth factors can have three distinct functions, with proliferative, anti-proliferative, and proliferation-independent effects. They are, in short, *multifunctional molecules* (Sporn and Roberts 1988).

In addition to growth factors, it has been found that many other molecules have multiple functions. Adrenaline, for example, is a neurotransmitter, but it is also a hormone produced by the adrenal glands to spring the body into action by increasing the blood pressure, speeding up the heart and releasing glucose from the liver. Acetylcholine is another common neurotransmitter in the brain, but it also act on the heart (where it induces relaxation), on skeletal muscles (where the result is contraction), and in the pancreas (which is made to secrete enzymes). Cholecystokinin is a peptide that acts as a hormone in the intestine, where it increases the bile flow during digestion, whereas in the nervous system is a neurotransmitter. Encephalins are sedatives in the brain, but in the digestive system are hormones which control the mechanical movements of food. Insulin is universally known for lowering the sugar levels in the blood, but it also controls fat metabolism and in other less known ways it is affecting almost every cell of the body.

The discovery of multifunctional molecules suggests that their function is not decided solely by their structure, but also by the *context* in which they find themselves. What matters, in other words, is not their ability to catalize a specific reaction, but the fact that they are employed as molecular signs that can be given one meaning in a certain context and a different meaning in another one.

A second finding that points to the existence of codes in signal integration is the fact that the regulation processes set in motion by second messengers are strongly conserved in evolution, and yet the actual reactions involved have undergone great changes in the history of life. The regulation of cellular energy homeostasis, for example, has been highly conserved from yeast to man, with the key role being played by a protein kinase that is called AMPK in animals and Snf1 in yeast. Despite this overall conservation, it has been found that an evolutionary divergence of about 150 million years between two species of budding yeasts (*Saccharomyces cerevisiae* and *Kluyveromyces lactis*) has produced substantial differences in their Snf1 regulatory networks (Charbon et al. 2004). Again, what seems to matter in these regulation processes is not a specific set of catalysts, but a set of rules that can be implemented in many different ways.

The information carried by first messengers, in conclusion, undergoes two great transformations in its journey towards the genes. First, it is transformed into internal messengers with the rules of the signal transduction codes, and then it is channelled along complex three-dimensional circuits that integrate it with other signals according to the rules of one or more *signal integration codes* (Barbieri 2003).

## 3.8 The Histone Code

The classic *double helix* described by Watson and Crick has a width of 2 nm (two millionths of a millimeter), but in eukaryotes many segments of this filament are folded around groups of eight histone proteins and form blocks, called *nucleosomes*, that give to the filament a ‘beads-on-a-string’ appearance. This string, called *chromatin*, is almost six times thicker than the double helix and is further folded into spirals of nucleosome groups, called *solenoids*, that arrange it in fibers of increasing thickness and ultimately into the 600 nm fiber of the *chromosome*.

These multiple foldings allow the eukaryotic cells to pack their long chromosomes into the tiny space of their nuclei, and for this reason it was initially assumed that the histones have a purely packaging role. The experimental data, however, have shown that the ‘tails’ of the histones (the parts that protrude from the surface of the nucleosomes) are subject to a wide variety of post-translational modifications (in particular acetylation, methylation and phosphorylation) that have highly dynamic roles and are involved in the activation or repression of gene activity (Kornberg and Lorch 1999; Wu and Grunstein 2000).

The histone tails represent about 25–30 % of the histone mass, and their post-translational modifications can alter the chromatin either directly or indirectly. The *direct* modifications are those that physically open or close the molecular space (in particular the electrostatic barrier) that surrounds the genes and in this way control the transit of DNA-binding proteins (Hansen et al. 1998; Wolfe and Hayes 1999). Several discoveries, however, have shown that the most frequent effects are obtained by *indirect* mechanisms. In these cases, the modified histone tails provide 'marks' on the surface of the nucleosomes that are recognized by specialized *effector proteins* which set in motion chains of biological reactions that eventually end in the activation or the repression of specific genes (Agalioti et al. 2002; Peterson and Laniel 2004; Berger 2007; Gräff and Mansuy 2008).

A crucial breakthrough, is this field, was the discovery that the post-translational modifications of the histones do not act individually. Most of them are involved in both the activation and the repression of genes (the phosphorilation of histone H3, for example, takes part in the condensation as well as in the decondensation of chromatin), which means that the final result is due to a *combination* of histone marks rather than a single one. This led David Allis and colleagues to propose that the histone marks operate in combinatorial groups, like letters that are put together into the words of a molecular 'language' that was referred to as *histone code* (Strahl and Allis 2000; Jenuwein and Allis 2001).

The same concept was independently proposed by Brian Turner (2000, 2002) who argued that there is an *epigenetic code* at the heart of the regulation mechanisms that are initiated by histone tail modifications. Turner pointed out that these modifications are *epigenetic* because they operate *in addition* to genetic changes, and underlined that they have both short-term and long-term effects. The short-term modifications change rapidly in response to external signals and represent a mechanism by which the genome quickly responds to the environment (Schreiber and Bernstein 2002). The long-term modifications, instead, are those that are put in place at early stages of embryonic development and allow the transcription or the silencing of specific genes at more advanced stages (Turner 2007).

The existence of long-term effects was revealed by the discovery that many histone modifications survive the trauma of mitosis and are transmitted to the daughter cells. This is particularly important in embryonic development where the cells must perpetuate their state of differentiation into distinct tissues. The histone modifications, in other words, provide a mechanism of *cell memory*, in the sense that they enable the cells to 'remember' their specific pattern of gene expression for many generations (Jeppesen 1997; Lyko and Paro 1999; Margueron et al. 2005; Xin et al. 2007). It has been shown, for example, that the expression of *Hox* genes in embryonic development is regulated by histone modifications (Zhu et al. 2005; Wang et al. 2007). Another example of long-term effects is provided by the histone modifications that allow neural cells to generate faster action potentials the more they are used, making the transmission of action potentials increasingly easier (Levenson and Sweatt 2005).

Today, in conclusion, a large number of data support the idea that the regulation of genetic activity by histone modifications plays a fundamental role in all eukaryotes and is based on the rules of a combinatorial code that has become known as 'histone code'.

## 3.9 Is the "Histone Code" an Organic Code?

This question is the title of a paper where Stefan Kühn and Jan-Hendrik Hofmeyr described the results of a research project dedicated to find out whether or not the histone code has *all* the essential characteristics of an organic code (Kühn and Hofmeyr 2014). The prototype example of the genetic code shows that an organic code requires three things: (1) two independent molecular worlds, (2) a set of molecular adaptors that create a mapping between them, and (3) the demonstration that the mapping is arbitrary because its rules can be changed. Kühn and Hofmeyr tested the histone code in respect to all these points.

1. *The Two Independent Worlds of the Histone Code*
   An organic code is a mapping between *organic signs* and *organic meanings*, and in many cases signs and meanings are both organic molecules. The genetic code, for example, is a mapping between codons and amino acids, whereas the signal transduction code is a mapping between first and second messengers. Kühn and Hofmeyr, however, pointed out that the organic meanings can be *biological effects* rather than molecules. In principle this may not seem an extension of the original definition because biological effects are necessarily implemented by molecules, but in practice it is a very useful generalization because there are cases in which a biological function is an experimental reality even when its molecular components are not fully known. And this is precisely the case in the histone code, where the organic signs are groups of histone modifications and the organic meanings are biological reactions that promote the activation or the repression of specific genes. The histone code, in other words, is a mapping between two independent worlds.
2. *The Adaptors of the Histone Code*
   The effector proteins of the histone code are the molecules that establish a bridge between organic signs and organic meanings, but in order to prove that they are true adaptors it is necessary to show that they operate independently on signs and meanings. Kühn and Hofmeyer underlined that this is precisely what happens because the effector proteins have two distinct domains: one that recognizes histone modifications and a different type that initiates biological reactions. It has been shown, for example, that the acetylated lysines are specifically recognized only by the bromodomains of the effector proteins (Owen et al. 2000; Mujtaba et al. 2007). The methylated amino acids are recognized by a greater variety

of domains but again each recognition step is absolutely specific (Cavalli and Paro 1998; Maurer-Stroh et al. 2003; Mellor 2006; Kim et al. 2006). The effector proteins, in other words, perform two independent recognition processes on signs and meanings and are therefore true adaptors.

3. *The Arbitrariness of the Histone Code*
   An organic code is arbitrary when its rules are not dictated by physical necessity and in this case it must be possible, at least in principle, to exchange the part of an adaptor that recognizes an organic sign with a different one and show that the modified adaptor associates the old organic meaning to the new sign. Kühn and Hofmeyr noticed that the experimental data support this possibility because there is evidence that the chromodomains of the effector proteins can be interchanged (Fischle et al. 2003).

   The histone code, in conclusion, did pass the three tests and Kühn and Hofmeyr ended their paper with these words: "Although we probably do not yet know the complete histone code, we have more than enough information to be able to recognize the histone code as a *bona fide* organic code."

## 3.10 The Tubulin Code

Tubulin is the major component of the microtubules, the filaments that form an internal scaffolding in all eukaryotic cells and give origin to organelles such as cilia, centrioles, basal bodies and the mitotic spindle. Most microtubules are in a state of rapid turnover by dynamic instability and alternate very quickly between growth and shrinkage. Within the cell, however, there is also a population of microtubules that are relatively stable, in the sense that their turnover is measured in hours rather than minutes. The function of the stable microtubules is still not completely known, but there are clear indications that they are involved in the morphogenesis of the eukaryotic cell (Kirschner and Mitchison 1986). What is certain, is that the stable microtubules undergo a variety of *post-translational modifications* (PTMs) that have been strongly conserved in evolution because they are found in all eukaryotic taxa.

These PTMs consist in processes like acetylation, phosphorylation, polyglutamylation, polyglycylation, detyrosination, and palmitoylation that act preferentially on stable microtubules. They have been studied with various tests on purified tubulin, but the experiments have failed to detect any direct effect of the PTMs on the dynamics of the microtubules (Maruta et al. 1986; Webster et al. 1990). This means that PTMs do not act by changing directly the intrinsic properties of the microtubules, but rather by providing combinatorial signals for the recruitment of proteins that interact with the microtubules. Different combinations of PTMs, in other words, act like signposts that specify the properties that stable microtubules are going to have in different regions of the cell or in different periods of the cell cycle. To this set of signposts that operate on stable microtubules, Kristen Verhey and Jacek Gaertig (2007) gave the name of *Tubulin code*.

Any organic code, as we have seen, requires molecules that act like adaptors between two different domains. Verhey and Gaertig have called these molecules '*interpreters*', and have identified three major classes of microtubule binding proteins that can be considered interpreters of the tubulin code: "First, microtubule associated proteins (MAPs) such as Tau, MAP1 and MAP2 that bind statically along the length of microtubules. Second, plus-end tracking proteins (+TIPs) that bind in a transient manner to the plus-ends of growing microtubules. And third, molecular motors that use the energy of ATP hydrolysis to carry cargoes along microtubule tracks."

Verhey and Gaertig have also called attention to a unique characteristic of the tubulin code. Many epigenetic modifications are transmitted from one generation to the next, but this does not usually happen in the tubulin world: "Some microtubule-based organelles (e.g., centrosomes and basal bodies) are inherited by a template-driven mechanism but there is no evidence that the template organelle directly influences the PTM pattern in the new organelle. Rather, the PTM pattern is recreated in the newly formed organelle in a gradual manner ... Other microtubule-based structures, such as cytoplasmic microtubules, the mitotic spindle and cilia, are formed *de novo* mostly, if not entirely, from unmodified tubulin heterodimers. Thus, in case of both template-dependent and template-independent microtubular structures, PTM patterns are probably recreated without a direct influence of preexisting PTMs."

The existence of the tubulin code, in conclusion, is based on sound experimental evidence but the actual deciphering of its rules is still at a preliminary stage and requires a detailed understanding of how the PTMs influence the recruitment of proteins and regulate the functions of the stable microtubules.

## 3.11 The Sugar Code

For a long time, sugars have been regarded as molecules that provide energy (mostly in the form of glucose and glycogen) or structural support (like cellulose in plants), but molecular biology has shown that they also have a third outstanding function: by binding to proteins they generate *glycoproteins*, molecules that take part in countless communication processes in and between cells.

The addition of sugars to proteins is a post-translational modification, called *glycosylation*, that greatly expands the potentialities of many protein families and gives origin to glycoproteins that perform a wide variety of functions. Some operate on the cell membrane and act as antennae for receiving molecular signals or as docking sites for importing compounds. Other glycoproteins take part in cell-to-cell interactions, for example in sperm-oocyte attachment, in bacteria-to-cell relationships and in the aggregation of platelets. A third family operates in the immune system where glycoproteins interact with antigens, recognize white blood cells, and take part in the *major histocompatibility complex* (MHC). Yet another family is that of the glycoproteins that act as hormones, like human corionic gonadotropin

(HCG), thyroid-stimulating hormone (TSH) and erythropoietin (EPO). Then there are glycoproteins that have protective functions (*mucins*), some that are involved in transport (*transferrin*) and others that act as enzymes (*alkaline phosphatase*).

The key point in these interactions is that in most cases it is the *sugar* component that determines the recognition ability of the glycoproteins. This point has been particularly underlined by Winterburn and Phelps (1972), who convincingly argued that "*the significance of the glycosyl residues is to impart a discrete recognitional role on the protein*". Sugars, in other words, are carriers of information because their sequences have specific biological functions, and yet the information they carry is only partially contained in the genome. In most cases it is due to subtle epigenetic modifications in the terminals of the sugar antennae (Gabius 2000).

It has been found, furthermore, that sugars have a capacity to store information that is many orders of magnitudes higher than that of nucleotides and amino acids (Laine 1997). This makes us realize that, after nucleotides and amino acids, sugars are a third great family of informational molecules, but how do they transmit their messages to the other components of the cell?

The key discovery, on this point, is that the functions that are associated with sugars are *not* set in motion by the sugars. In most cases, they are set in motion by proteins that interact with the sugars and recognize the specific role that they have in any given set of circumstances. These *sugar-binding proteins* became popular in the early 1900s mainly because they served to determine the chemical structure of the ABO blood groups and were originally called *agglutinins*. In 1954, however, Boyd argued that they should be given a new name that reflects the unique function that they actually perform, i.e., the highly specific selection of carbohydrates. To this purpose he proposed to call them *lectins*, on the ground that this term derives "from the Latin *lectus*, the past principle of *legere* meaning to pick, choose or select" (Boyd 1954).

The next step in the discovery of the informational properties of the sugars was the recognition, by Hans-Joachim Gabius, that their messages must be *decoded* in order to have biological effects, and that lectins are the decoding devices in this process. Gabius, in other words, realized that lectins are *adaptors*, molecules that act as intermediaries between sugars and biological reactions and establish connections between them that are not determined by physical necessity. This is why he proposed that there is a *Sugar code* at the basis of the communication processes that involve sugars, and that "*lectins are the translators of the Sugar code*" (Gabius 2000, 2009).

## 3.12 The Glycomic Code

An extracellular matrix called *cell wall* surrounds all plant cells and one of its most common component is *cellulose*, a polymer formed by long chains of glucose that bind to each other with such great affinity that most of the water is excluded from their surface. The result is a structure that is very hard to hydrate and to break. Cell walls, however, are not made of cellulose only. There are other polymers that occur in significant amounts and in most cases they are similar to cellulose in structure, but are branched in more complex ways. Because of their similarity to cellulose, these branched polysaccharides have been called *hemicelluloses*. They surround the cellulose microfibrils and interact with each other with non-covalent bonds in such a complex way that the hemicelluloses are even harder to disassemble. On top of that, there is an even higher level of complexity: the cellulose-hemicellulose domain is embedded in a matrix of *pectins* (polysaccharides with very complex chemical structure) which forms a jelly-like structure that retains water and at the same time it further reduces the pores of the cell wall.

The evolution of this complexity is probably due to the fact that the cell walls, in addition to controlling the expansion and the growth of the plant cells, must also form a barrier that prevents, or makes it extremely difficult for, microorganisms to enter into the cell cytoplasm.

When microorganisms invade a plant cell, it may seem that all they need to gain access is a few enzymes that degrade pectins, hemicellulose and cellulose, but that is not the case. In fact, if microorganisms could easily enter into plant cells, most plants would not survive and life on Earth would not exist in its present form. So, how did plants manage to defend themselves?

The second most abundant polysaccharide on Earth after cellulose, is *xyloglucan*, and by using enzymes such as *cellulases*, researchers could study whether the oligosaccharides found in xyloglucan were arranged randomly or not. A first answer to this question came from the discovery by Buckeridge et al. (1997) of a new xyloglucan polymer that contains two families of oligosaccharides, one with four and the other with five glucoses (tetramers and pentamers). In 2006 Marco Tiné and Marcos Buckeridge, working with Nick Carpita from Purdue University, demonstrated with mass spectrometry that there are regularities in the tetramers and pentamers of the xyloglucan molecules (Tiné et al. 2006). This was probably the first proof that the constitutive blocks of xyloglucan are not arranged randomly.

After that finding, Marcos Buckeridge and Amanda De Souza performed experiments on a large number of hemicelluloses and found that some enzymes have higher specificity for certain regions in all branched polymers, which implies that their molecules too are non-randomly organized. These regularities in hemicelluloses suggested that their assembly is controlled by rules, and the fact that they are the result of contingent evolutionary developments indicated that the rules are arbitrary. This is why the authors proposed that there is a *glycomic code* in plant cell wall hemicelluloses (Buckeridge and de Souza 2014).

A consequence of this proposal is the idea that plants evolved an increasingly complex system of coding rules for the assembly of hemicelluloses in order to keep at bay the invading organisms by forcing them to develop an increasingly high number of specific enzymes. As a result, only a few microorganisms managed to find the *key* to enter any given plant cell. The glycomic code, in other words, was probably the result of a co-evolutionary race between plants and herbivores.

The constraints imposed by the glycomic code on plant cell walls are so severe that many organisms – including us – have evolved digestive systems that are totally dependent on cell walls (familiarly known as *food fibers*). Furthermore, plants themselves hardly degrade their own walls. It is true that in a forest cell walls are eventually degraded, but this is achieved by *communities* of microorganisms and never (or rarely) by a single species. If the glycomic code of plant cell walls did not exist, in conclusion, we would probably not be here because plants would be utterly different from what they presently are.

## 3.13 The Sequence Codes

In the 1980s and 1990s, Edward Trifonov proposed that genomes carry several codes simultaneously, not just the classic genetic code, and gave them the collective name of *sequence codes*. This conclusion rests upon Trifonov's definition that *"a code is any sequence pattern that can have a biological function"* or *"a code is any pattern in a sequence which corresponds to one or another specific biological function"* (Trifonov 1989, 1996, 1999).

The concept that a '*code is a sequence*' is at variance with the traditional idea that '*a code is a mapping*', but does not exclude it, and this suggests that codes can be defined in two different ways. This, after all, is not uncommon. Information, for example, has been defined by Watson and Crick as a *linear sequence* and by Shannon as a *probability function*. In communication theory, furthermore, the terms *source code* and *channel code* are sequence codes, not mappings, and represent a well established tradition. It seems therefore that organic codes too can be defined in different ways, and all we need to keep in mind is that two distinct definitions are valid when they represent two distinct aspects of reality.

With this understanding, the idea of the sequence codes is the acknowledgement that various interlocking messages do exist in genome sequences. One of them, for example, is the message that carries information about where nucleosomes should be positioned on DNA, a message to which various authors have given the name of *nucleosome positioning code* (Yuan et al. 2005; Segal et al. 2006).

A second example, is the fact that various aspects of chromatin folding appear to be pre-programmed, and the folding rules that are encoded in sequences have been referred to as *chromatine folding code* (Boutanaev et al. 2005). A third example comes from the discovery that in many proteins the choice of codons is highly biased and the preference for particular codons is often determined by factors that are not related to protein structure or function. Hence the idea that the choice of codons is subject to additional constraints that have been referred to as *regulatory codes* of chromatin organization, or, more precisely, as *transcription factor binding regulatory codes* (Stergachis et al. 2013; Wheatheritt and Babu 2013).

Various other sequence codes have been reported in the literature, but their best classification is still the one proposed by Trifonov, who suggested that there are at least eight families of sequence codes in addition to the classic triplet code (Trifonov 1996). Here is how he described them:

1. The *transcription codes* include promoters and terminators, and are rather universal, though different in prokaryotes and in eukaryotes.
2. The *gene splicing code* for the processing of nuclear pre-mRNA is largely undeciphered. Its main components are obligatory GU- and AG-ends of introns, as well as rather conserved consensus sequence features around the ends.
3. The *translation pausing code*, for the regulation of translation, is encoded by clusters of rare triplets for which the aminoacyl-tRNAs are in limited supply.
4. The *DNA structure code*, or *DNA shape code*, is a sequence-dependent local shape of DNA which is a crucial component of the protein-DNA recognition.
5. The *chromatin code* describes those sequence features that direct the histone octamer's binding to DNA and the formation of nucleosomes.
6. The *translation framing code* is overlapping with the triplet code, and ensures the correct reading frame during translation.
7. The *modulation code* is about the repeating sequences and regulates the number of repeats as an adjustable variable to modulate expression of the nearby gene.
8. The *genome segmentation code* is one of the emerging new codes, and is due to fact that the genomes appear to be built of rather standard size units.

## 3.14 Organic Codes and Biology

The discovery of the genetic code was immediately recognized as one of the greatest scientific revolutions of all times. One of the best documents of its extraordinary impact is the fact that the Nobel price for Medicine or Physiology was awarded to Niremberg, Khorana and Holley in 1968 – only 2 years after the complete deciphering of the code. The popularity of genetic code, furthermore, spread far beyond the scientific community and fired the imagination of many other intellectuals and eventually of the public at large.

This is in complete contrast with what happened a few decades later to the other organic codes, whose discoveries circulated only in small groups and were virtually ignored by most biologists. There may be many reasons for this, but one of them is probably the fact that the first models of the genetic code were all inspired to the *stereochemical hypothesis*, the idea that the coding rules are the result of chemical interactions in space. If the genetic code can be reduced to chemistry, the same would apply to all other organic codes, and their study would be the concern of a few specialized areas of biochemistry.

One could argue that the genetic code is a real mapping, based on real adaptors, between two worlds of molecular sequences, and sequences involve information whereas mappings involves meaning, but it was objected that these new concepts do not belong to biology. They can be left to humanistic disciplines like epistemology and philosophy. There is however one aspect of the organic codes that cannot be excluded from biology, and that is their implications for the history of life.

One of the best example of the importance of history in biology was the rise of *cladistics*, the science created by Willi Hennig in the 1950s and 1960s that revolutionized the study of phylogenesis. Hennig was able to show that population genetics, adaptation and natural selection have little to do with the reconstruction of genealogical trees, and proved that historical relationships can be obtained from strictly logical and scientific arguments. What Hennig demonstrated, in other words, is that the reconstruction of the history of life *with scientific methods* is possible and is an integral part of biology.

Something similar applies to the organic codes. Biology can avoid dealing with their epistemological and philosophical issues, but cannot avoid dealing with their *history*. If many organic codes exist today, it is because they had origins and evolutions, and this is an issue that goes at the very heart of biology.

The rest of the book is dedicated precisely to this goal: to the study of all codes of life, to a first tentative reconstruction of their history, however imperfect that may be, and finally to the transformation of the code paradigm into *code biology*.

# Part II
# Major Steps in Macroevolution

# Chapter 4
# Life Before the Cell

Ever since the classic experiment by Stanley Miller (1953) it is known that many organic molecules can be formed spontaneously in a variety of environmental conditions. Organic molecules have been found, for example, in meteorites that arrived on Earth, and spectroscopy has revealed their presence in comets and in interstellar space. This tells us that organic matter is formed spontaneously whenever and wherever suitable conditions exist, and today we know that those conditions existed on Earth at a very early stage of its history.

For more than a century, now, the problem of the origin of life has been formulated as the problem of describing the origin of the first cells on the primitive Earth. More precisely, as the problem of describing how primitive genes and primitive proteins appeared on Earth and managed to evolve into supramolecular systems that had the characteristics of primitive cells.

The crucial point in this process is to realize that genes and proteins are *not* produced by spontaneous reactions in any cell. They are produced by molecular machines that physically stick their subunits together according to the order provided by templates. Genes and proteins, in short, are *manufactured* molecules, not spontaneous ones. Spontaneous genes and spontaneous proteins did appear on the primitive Earth but could not evolve into cells because they did not have biological specificity. They evolved instead into molecular machines and it was these machines and their products that gave origin to the first cells.

The distinction between spontaneous and manufactured molecules allows us to divide the period that preceded the first cells into two great phases: a first period of *chemical* evolution during which organic compounds were formed exclusively by spontaneous reactions, and a second period of *postchemical* evolution, that started with the appearance of the first molecular machines and of the first manufactured molecules. Postchemical evolution, in short, is a stage of the history of life that came after chemical evolution but had not yet acquired the full characteristics of biological evolution.

Unfortunately, postchemical evolution is virtually an unexplored field of research, today, and is largely based on speculations, so it is tempting to dismiss it.

© Springer International Publishing Switzerland 2015

M. Barbieri, *Code Biology*, DOI 10.1007/978-3-319-14535-8_4

But speculation on the origin of the cell is not a useless exercise because we understand something only when we understand, at least in principle, how it came into being. That is why origins are so important and why we are so fascinated by them.

## 4.1 The Twin Problems of the Beginning

The simplest molecular machines that could appear spontaneously on the primitive Earth were *bondmakers*, molecules that could stick monomers together in a random order and produce statistical polymers. In a primitive environment where chemical evolution had already accumulated many varieties of organic molecules, bondmakers could form chemical bonds at random between most of them, thus giving origin to polymers such as polypeptides, polynucleotides, polysaccharides and countless other combinations of their building blocks.

Random polymers could also be formed without bondmakers, by the spontaneous aggregation of monomers, but bondmakers had at least two important advantages. One is that they could produce polymers at a virtually continuous rate, thus enormously increasing their number on the primitive Earth. The other is that some bondmakers could acquire the ability to join monomers together no longer at random but in the order provided by templates. These bondmakers, in short, started making copies of the templates and became *copymakers*. The appearance of bondmakers led therefore to a steady increase in the production of random polymers and eventually to the appearance of the first copymakers, the first molecular machines that started populating the Earth with a potentially unlimited number of copied molecules, in particular of nucleic acids.

Today, the copying of nucleic acids is performed by large molecules called *polymerases* but the formation of a chemical bond does not necessarily require big dimensions and it is likely therefore that the first polymerases were smaller than their modern descendants. What really matters, for our purposes, is that the origin of molecular copying does not seem to require extremely improbable events. At an advanced enough stage of chemical evolution, in an environment that had already produced a wide variety of organic molecules, including nucleotides and nucleic acids, the appearance of copymakers was as likely as that of any other average-size molecular structure.

The evolutionary impact of that event, however, was enormous, because the copying of nucleic acids leads in the short run to heredity and in the long run to natural selection. Molecular copying, in short, gives origin to a chain of processes that is potentially unlimited and endlessly open to change. Heredity and natural selection, the very hallmarks of life, are the result of a copying mechanism that is extraordinarily simple.

But there is more to life than heredity. There is metabolism, the actual building and re-building of organic structures and the never ending exchange of energy and matter that keeps everything going. And here things are much more complicated, because the molecular agents of metabolism are proteins and proteins cannot be copied. They are manufactured by huge molecular machines, called *ribosomes*, according to the rules of the genetic code, and even the simplest apparatus of protein synthesis requires the coordinated contribution of more than 100 different molecular components.

The origin of genes, in short, can be accounted for by a relatively simple mechanism, but the origin of proteins demands an apparatus whose complexity is almost beyond imagination. These are the twin problems of the beginning, and the second is probably the hardest of all the mysteries that we need to solve before we can say that we understand, at least in principle, how life appeared on our planet.

## 4.2 The Replication Paradigm

An unexpected solution to the problem of the origin of metabolism came in the 1980s with the discovery that some RNAs can function not only as genes but also as enzymes (Cech 1983, 1986; Guerrier-Takada et al. 1983, Guerrier-Takada and Altman 1984). This suggested that RNAs capable of enzymatic activity, called *ribozymes*, could well have been the first agents of metabolism in the initial stages of the history of life. As Walter Gilbert (1986) put it: "*There is no need for protein enzymes at the beginning of evolution. One can contemplate an RNA world containing only RNA molecules that serve to catalyse the synthesis of themselves*".

The idea that the first living systems were made of RNAs became a realistic possibility, and this gave a great impetus to the *replication paradigm*, the idea that life has been the result of a process of chemical evolution based on self-replicating molecules. This idea is still popular, today, but it must be underlined that it has been tested in the laboratory by two complementary experiments performed by Sol Spiegelman (1967) and by Manfred Eigen (1971), and the results have not supported it.

Spiegelman studied the evolution in the test tube of long RNA molecules obtained from a virus. Initially, these molecules contained about 4,500 nucleotides that were coding for a complete set of viral proteins, but in the course of the experiment shorter mutants started appearing by the chance loss of genes that were not essential to replication, and in the end all that remained in the test tube was a relic of 220 nucleotides that could be replicated indefinitely (the *Spiegelman monster*). Eigen, on the other hand, studied the evolution of RNAs starting from the opposite end, i.e., from free nucleotides in a solution that contained also an enzyme that could join them together. In these conditions, the nucleotides were assembled into molecules that grew longer but only up to a limit, and in the end the solution reached a stationary state where the molecules capable of indefinite replication were, on average, 120 nucleotides long.

The two experiments arrived in this way at the same conclusion: the evolution of self-replicating RNAs does not produce an unlimited increase because there is a severe limit to the size of the genes that are capable of indefinite replication. But how general is this conclusion? Can we extend it from the test tube to real life?

Eigen and Schuster (1977) showed that the limit in question is indeed universal because it is the result of fundamental theorems that apply to all self-replicating systems. The maximum length of the molecules is determined by the replication errors that are inevitably present in any replication process, because beyond that limit the system is overtaken by a runaway *error catastrophe* and collapses. Eigen and Schuster proposed that the error catastrophe could be avoided if different molecules combine their replication cycles into a single *hypercycle*, but Ursula Niesert was able to prove that hypercycles too have a size limit beyond which they are swept away by three new types of error catastrophes (Niesert et al. 1981).

Since then, the problem of the maximum size that is attainable by self-replicating molecules has been studied under a variety of conditions, with or without compartments and with or without the arrangement of genes in chromosomes, but in all cases it has not been possible to avoid error catastrophes (Maynard-Smith and Szathmáry 1995). All theoretical and experimental results collected so far, in other words, show that the evolution of self-replicating molecules can produce little more than short sequences, and nothing remotely similar to a small cell.

The replication paradigm, in short, does not offer a realistic model for the origin of life. This means that precellular evolution had to take place in systems that did not depend on exact replication. More precisely, in systems that had the ability to tolerate errors within very wide limits, because only such systems could avoid the error catastrophes.

## 4.3 Ribosoids

Molecular phylogeny has shown that the ribosomal RNAs are among the most conserved molecules in evolution (Woese 1987, 2000) and this tells us that they appeared very early on Earth. The first RNAs, furthermore, appeared together with countless other molecules, and started interacting with them. Of all such interactions, those between RNAs and peptides were particularly important because the functions of most RNAs are greatly enhanced by the attachment of peptides, even small ones (Orgel 1973). This is why all molecules formed by RNAs and by combinations of RNAs with peptides and proteins (ribopeptides and ribonucleo-proteins) have been collectively referred to as *ribosoids*, and the population of all ribosoids in a system has been called the *ribotype* of that system (Barbieri 1981).

The theories on the origin of life have been traditionally divided into *gene-first* and *protein-first* models, but the ribosoids suggest a third possibility. In addition to the gene-first and to the protein-first scenarios, we can consider the idea that genes and proteins appeared and evolved *together* on the primitive Earth. A particular version of this idea is *the ribotype theory on the origin of life* (Barbieri 1981, 1985), a theory that here is re-proposed in an expanded and updated form.

The chemical properties of RNAs and peptides were undoubtedly favouring their interactions, but that is not enough to account for a long lasting *partnership*. Why then did RNAs and peptides evolve together?

An important clue has come from the discovery that short pieces of ribosomal RNA can form peptide bonds between amino acids, thus joining them together and producing polypeptides (Nitta et al. 1998). This means that some primitive RNAs could produce random strings of peptides, and the interaction of these peptides with other RNAs could allow them to perform a variety of functions, including the ability to form phosphodiester bonds between nucleotides, thus joining them together and producing polynucleotides.

Some ribosoids, in short, became *polimerizing ribosoids* and started producing random polypeptides and random polynucleotides. There is no biological specificity in these relationships, and yet the mechanism for a sustained partnership is already there. *RNAs and peptides evolved together because one class of molecules provided bondmakers for the other.*

The polimerizing ribosoids that were sticking amino acids together were the ancestors of the ribosomes, and it is likely that were formed by self-assembly from their components because modern ribosomes still have that property. It must be underlined, furthermore, that the polymerizing ribosoids did not need to have *identical* structures in order to perform the same function. Ribosomes of different species have different proteins and yet they all translate messenger-RNAs equally well, as if they were identical, which means that the ability to polymerize amino acids is shared by many different configurations of ribosoids.

This allows us to conclude that a statistical form of replication was possible long before the origin of true replicating systems. This statistical mechanism has been referred to as *quasi-replication* (Barbieri 1981) because the descendants were not exact copies of the progenitors. The peptide-polymerizing ribosoids, or *protoribosomes*, were capable of quasi-replication because they could produce a wide variety of random proteins some of which could interact with ribosomal RNAs and self-assemble into new protoribosomes which had the same function of their predecessors even if their molecules were different.

A quasi-replicating system has three main characteristics. The first is that only a fraction of its products are reinvested in the reproduction of the system (*enormous waste*). The second is that the original system is reproduced by self-assembly of its individual components (*no transfer of information*). The third is that the components which reproduce the system need not be exact copies of the components of the original system (*polymorphism*). This third property was particularly important because it was allowing the systems to increase their size and complexity without being swept away by error catastrophes.

## 4.4 Nucleosoids

Ribonucleic acids and proteins can form not only ribonucleoproteins and assemblies of ribonucleoproteins like ribosomes. They can also generate giant scaffoldings made of tens of thousands of molecules and produce supramolecular systems whose dimensions can reach the size of a small cell. The formation of these supramolecular clusters is based on processes of self-assembly, so it could well have taken place in primeval solutions, particularly when these became enriched by RNA-driven synthesis of proteins and by protein-driven synthesis of RNAs. The limits to the dimensions of such clusters are anybody's guess, but the best example that we have today is represented by the nucleoli, and for this reason they have been given the name of *nucleosoids* (Barbieri 1981).

Nucleosoids are therefore clusters of ribosoids, and form a highly heterogeneous family whose members can have shapes and dimensions of countless different types. Most of the nucleosoids that appeared on the primitive Earth were bound to be dead-ends from the point of view of evolution, but some of them could have an evolutionary potential. The comparison with modern nucleoli cannot be taken literally, of course, but it is nonetheless useful because it allows us to discuss some general properties that can be attributed to all nucleosoids.

1. The first property is that a cluster of ribosoids is by no means an inert scaffolding. For one thing, its ribonucleoproteins can perform physical movements from one point in space to another and undergo a variety of conformational changes despite their being part of a supramolecular structure.
2. Secondly, clusters of ribosoids provide microenvironments which trap molecules and localise their interactions. This is particularly important when many components take part in a biological process because molecules need to be very close to each other in order to interact, and micro-compartments are ideally suited to keep them at short distances.
3. Thirdly, nucleosoids form three-dimensional backbones which can have all intermediate dimensions between the nanometer and the micron, thus providing a bridge between molecules and cells.

Internal movements, internal compartments and a wide range of dimensions were properties of many types of nucleosoids and gave them a distinct evolutionary potential. It may be pointed out that nothing is of value in evolution if it has no lasting effect and however interesting the nucleosoids were, they were not capable of replication. The answer, once again, is *quasi-replication*.

We know that nucleoli of different species have widely different sizes, shapes and components and yet they all perform the same function as if they were identical (Busch and Smetana 1970). It is perfectly legitimate to assume that a similar polymorphism was present in primitive nucleosoids, and we can assume therefore that they too were capable of quasi-replication.

## 4.5 Nucleosoid Evolution

Nucleosoids could synthesize a wide variety of compounds and their supramolecular organization could trap the ribosoids in a confined space, so that the cycles of synthesis could go on for long periods of time. Many nucleosoids were presumably making components that did not have a future, but on purely statistical grounds we can assume that a fraction of them were preferentially synthesizing other ribosoids and these did have an evolutionary potential. A nucleosoid that was able to synthesize its own components could *grow* and reach whatever dimensions were physically attainable. Eventually, however, it would become unstable, break apart in smaller pieces and in some of these the ribosoids that were responsible for the previous syntheses would go on repeating the process.

It is worth underlining that the breaking up of nucleosoids would distribute its fragments in space, and among these there were pieces of RNAs that could travel and colonize other nucleosoids, thus behaving like the viruses that are responsible for what has become known as *horizontal gene transfer* (Doolittle 1999; Woese 2000). As in the case of the ribosoids, the quasi-replication of the nucleosoids was based on the fact that the descendants do not have to be exact copies of the progenitors.

The growth and breaking up of the nucleosoids were determined by the chemical interactions between their molecules and could vary within very wide limits. Not all combinations, however, were destined to leave viable descendants. Some nucleosoids, for example, could go on growing almost indefinitely and were probably condemned to choke to death. Others were breaking up with extreme ease and in that case they were likely to produce descendants that missed some essential components of their syntheses.

In order to leave viable descendants, nucleosoids had to strike a balance between growth and break-up, and it is likely that only a few of them managed to achieve it. It is also possible, however, that some parallel events came to the rescue and improved their long term survival. The ability to split up into two fairly similar parts would have been an advantage for the descendants, and would have been highly favoured by the appearance of some structural guideline for splitting.

As it happens, we do have a plausible solution to that problem. DNA is a molecule that can be split into two complementary halves, and each of them can provide attachment sites for countless types of ribosoids. DNA was likely to exist in primitive solutions because it can be formed from RNA by removing one oxygen from the ribose, and it was ideally suited for splitting a cluster of ribosoids into fairly similar halves. In that case DNA did not have to behave as a carrier of information and was recruited only for structural purposes, which explains why it could have appeared at a stage when biological specificity did not yet exist. It is possible, in short, that DNA started functioning as a structural guideline for the splitting of nucleosoids, and this allowed them to produce descendants for a longer number of generations.

The evolution of nucleosoids, however, could not go on indefinitely. Their quasi replication mechanisms were producing an unpredictable variety of random components and were bound to lead them either towards extinction or towards yet more complex structures.

## 4.6 Heterosoids

The nucleosoids are clusters of ribosoids, but their statistical composition implies that many other types of molecules could become associated with them. As long as these non-ribosoidal components had a temporary association with the nucleosoids, their presence was a sort of random noise and can be ignored. Eventually, however, some 'contaminations' proved to be useful for quasi-replication purposes, and that turned them into increasingly stable components. The nucleosoids evolved in this way into structures that became tightly associated with non-ribosoidal components, a process that can be summarized by saying that the nucleosoids became *heterogeneous nucleosoids* or *heterosoids* (Barbieri 1981).

One such contamination was DNA. The natural tendency of DNAs to split into two complementary parts could have greatly helped the splitting of the nucleosoids into fairly similar pieces, and this can explain why DNA became a component of quasi-replicating systems even when it was not yet carrying genetic information.

In addition to DNA, there was at least another class of molecules that had the potential of becoming permanently associated with nucleosoids. The research in prebiotic chemistry has shown that lipids and lipid-like molecules can be formed fairly easily in simulation experiments and are frequently present in meteorites. This means that they are high-probability molecules, and it is likely therefore that they were formed in great quantities on the primitive Earth. Lipids, on the other hand, are capable of spontaneous self-assembly into supramolecular structures such as vesicles and membranes, and lipid membranes could easily reappear in different generations because they have a sort of self-reproducing potential of their own. Most membranes in fact are generated from other membranes, and this means that they do not have to be created *ex novo* at each generation (Luisi and Varela 1989; Morowitz 1992; Szathmáry 1999).

One of the most important characteristic of the lipid membranes is their ability to act as catalysts in a variety of reactions such as ion transport and signal transduction. In particular, they can incorporate pigment molecules that capture the energy of light and set in motion those chains of energy transduction processes that eventually led to photosynthesis and respiration.

Lipid membranes, in short, could provide the sites of energy transformations and this was extremely valuable in RNA-based systems because RNAs have no ability to capture energy from the environment. As a matter of fact, the capacity to act as energy sources was so important that the chemical evolution of lipid systems was probably a precondition for the evolution of ribosoids and nucleosoids. It is possible, in other words, as suggested by Segré et al. (2001) that "*a Lipid World may have preceded the RNA world*".

There have been, in other words, two major developments in the evolution of the nucleosoids: DNA molecules started appearing inside them, and lipid membranes started providing energy sources and protective coats around them. These processes of integration gave origin to the first *modular structures*, i.e., to systems made of subsystems that were performing different functions but managed to form a single working whole.

The first heterosoids, however, were not yet the first cells, because they were still synthesizing statistical proteins and producing unequal descendants. In order to become cells they had to be able to transmit specific proteins to their descendants, and this required an apparatus that was making proteins no longer at random but according to the rules of a code.

## 4.7 A Primitive Apparatus

Modern ribosomes consist of a small and a large subunit that are made of ribosomal-RNAs (rRNAs) and ribosomal proteins (r-proteins). The ribosomal-RNAs account for more than half of the molecular weight of the ribosome, and the rest is accounted for by more than 50 different ribosomal proteins of low molecular weight, each present in one or a few copies (Nomura et al. 1974).

In order to make proteins, modern ribosomes interact with two other types of RNAs: a messenger-RNA (mRNA) that carries genetic information in the form of a sequence of codons, and a set of transfer-RNAs (tRNAs) that perform two distinct operations. At one end they recognize the codons with complementary anticodons, and at the other end they transport the amino acids that are added to the growing protein chain.

The mRNA attaches itself to the small subunit, whereas the formation of the peptide bonds occurs in the large subunit, at the *peptidyl transferase center*. The small subunit is therefore the decoding place of the ribosome, whereas the large subunit is the place of protein synthesis. The tRNAs create a bridge between them because they interact with the mRNA in the small subunit and deliver their amino acids to the peptidyl centre in the large subunit.

Various experimental results (Ban et al. 2000; Wimberly et al. 2000; Hsiao et al. 2009), have proved that ribosomal proteins are *not* present in the peptidyl center and this confirms that the ribosome is fundamentally an RNA machine. The ribosomal-RNA of the large subunit is a huge molecule (in bacteria it is 1,600 nucleotides long and has a molecular weight of one million), but its peptidyl transferase center is less than 100 nucleotides long, so it is likely that the ancestral ribosomal-RNAs were considerably smaller than their modern descendants.

These characteristics of the modern apparatus of protein synthesis allow us to reconstruct at least some of the characteristics of its ancestral precursors.

*Ancestral Ribosomes*

The ribosomal RNAs are among the most conserved molecules in evolution, and this suggests that the most primitive ribosomes were pieces of ribosomal-RNAs, probably stabilized by small polypeptides, that had the ability to form peptide bonds (Nitta et al. 1998) and were producing random chains of amino acids, i.e., statistical proteins.

*Ancestral Transfer-RNAs*
All modern transfer-RNAs are small molecules (75–90 nucleotides long) with a basic cloverleaf structure that has been highly conserved in evolution, a clear proof that they descended from a common ancestor. It is likely that they were recruited because of their ability to deliver amino acids to the site of synthesis, but it must be underlined that their contribution required the presence of a third type of RNAs.

*Anchoring-RNAs*
When the amino acids are delivered to the site of protein synthesis, it is necessary that they are kept at a close distance for a long enough time to allow the formation of a peptide bond (Wolf and Koonin 2007; Fox 2010). This means that the transfer-RNAs must have temporary anchoring sites, and in primitive systems these were provided by *anchoring RNAs*, the ancestors of the modern *messenger-RNAs*.

The combination of ribosomal-RNAs, transfer-RNAs and anchoring-RNAs gave origin to an apparatus of protein synthesis that presumably was more efficient than one formed exclusively by ribosomal-RNAs. Even this improved apparatus, however, could only produce statistical proteins and this gives us a problem: what was the good of these proteins? How could they favour evolution in the ancestral supramolecular systems?

## 4.8 Statistical Proteins

Amino acids are divided into four great groups (acidic, basic, hydrophilic and hydrophobic) and the chemical behaviour of proteins is largely determined by the combinations of these four types of molecules. Histidine and lysine, for example, are basic amino acids, and increasing their number would automatically produce more basic proteins whatever are their sequences. Chains of amino acids, furthermore, have a tendency to form either helices or pleated sheet (zigzag structures) and these configurations are strongly influenced by the chemical features of their amino acids. Helices, for example, are rich in alanine, leucine and glutamic acid, whereas zigzags are preferentially rich in methionine, valine and isoleucine.

Another important feature in proteins is the presence of hydrophilic and hydrophobic regions. The hydrophilic amino acids favour interactions in an aqueous medium whereas the hydrophobic ones allow biochemical reactions to take place in poorly hydrated environments.

Statistical proteins lack specific sequences and for this reason cannot act as specific enzymes, but can still perform many useful functions. In precellular systems, to start with, the RNAs were operating both as genes and enzymes but could barely work on their own. They needed peptides and polypeptides to maintain stable conformations in space or to have attachment sites for their interactions, and these *ancillary* functions required only generic chemical features not specific sequences.

Statistical proteins, furthermore, could arise from different combinations of acidic, basic, hydrophilic and hydrophobic amino acids, and even without specific sequences this gave them the ability to form helices or pleated sheet, to separate hydrophilic from hydrophobic regions, to assemble themselves into supramolecular structures, to create microenvironments and above all to maintain the system in a state of continuous metabolic activity. In primitive systems that were based on statistical proteins, in short, a fraction of these were likely to have useful chemical features. But could these systems evolve?

The key point, here, is that the probability that useful features appear in a population of statistical proteins depends upon the number of differences that exist among them, and this in turn depends upon the number of *different amino acids* that are used in protein synthesis. If only one amino acid is employed, the sole difference between the resulting proteins is their length; with two amino acids some chemical diversities start appearing, with three amino acids the diversities are more numerous, and so on.

The ancestral systems, in conclusion, could evolve only by introducing more diversity in their statistical proteins, and could do this only by increasing the number of amino acids in protein synthesis. But do we have any evidence that such increase took place in the early days of evolution?

The 20 canonical amino acids of the genetic code are divided into two great categories: less than half can be synthesized from inorganic molecules and for this reason are called *primary* (or *precursor*) amino acids. The others are synthesized by taking one of the primary amino acids as starting point and are referred to as *secondary* (or *product*) amino acids. This distinction is based on the biochemical syntheses that we observe today, but when a few other experimental facts are taken into account it acquires a deep evolutionary meaning. It has been found that secondary amino acids are never formed in laboratory experiments that simulate prebiotic conditions, and this strongly suggests that *only primary amino acids* appeared on the primitive Earth (Wong and Bronskill 1979; Wong 1981; Di Giulio 2008). This conclusion is reinforced by the discovery that the amino acids that are missing in laboratory syntheses are also missing from meteorites (Higgs and Pudritz 2007).

The very existence of secondary amino acids, in other words, implies that the number of amino acids employed in protein synthesis did increase in evolution: it started with less than 10 primary amino acids and steadily went up by the addition of secondary amino acids until it reached the canonical set of 20. Primitive systems based on statistical proteins, in conclusion, could evolve by introducing more amino acids in protein synthesis, and we know that such increase did take place in the early history of life. Our problem, at this point, is understanding the further evolution of those systems.

## 4.9 First and Last Common Ancestor

The modern genetic code is a mapping between 64 codons carried by transfer-RNAs and 20 amino acids carried by 20 aminoacyl-tRNA-synthetases, each of which attaches one amino acid to one or more tRNAs. The synthetases, on the other hand, are proteins that can be made only by an apparatus that already has a genetic code, and this gives us a classic *chicken-and-egg* paradox: how could the genetic code come into existence if its rules are implemented by proteins that can be produced only when the code already exists?

The logical solution of this paradox is that the *modern* apparatus of protein synthesis was preceded by an *ancient* apparatus where the amino acids were attached to the transfer-RNAs not by proteins but by RNAs, as suggested, for example, by Maizels and Weiner (1987). This amounts to saying that the *modern genetic code* based on protein-synthetases was preceded by an *ancient genetic code* based on RNA intermediaries. But how did the ancient genetic code came into existence?

The appearance of ribosomal-RNAs, transfer-RNAs and anchoring-RNAs made it possible to assemble an apparatus where the transfer-RNAs were delivering amino acids to ribosomal-RNAs in the order provided by the codons of the anchoring RNAs. This apparatus was automatically creating a mapping between codons and amino acids, and any such mapping is, by definition, a *genetic code*. We realize in this way that the first genetic code appeared on Earth when the first transfer-RNAs and anchoring-RNAs became involved in protein synthesis. We have seen, on the other hand, that the number of amino acids in protein synthesis increased in the early history of life, and the addition of any of them amounted to a change in coding rules. The first genetic code underwent therefore a number of changes, and gave origin in this way to *a family* of codes that here is collectively referred to as the *ancestral genetic code*.

The key point that must be underlined, in this case, is that a genetic code can be *ambiguous*. If a codon is allowed to code for two or more amino acids, a sequence of codons would be translated some time into a protein and some other time into a different protein, and the apparatus would produce *statistical* proteins, not specific ones. We know that the modern genetic code is *non-ambiguous* because every codon codes for one and only one amino acid, but what about the ancestral code?

In this case, the evidence suggests that ambiguity was bound to be present (Fitch and Upper 1987; Osawa 1995), and we can easily understand why. The most primitive ribosomes were probably pieces of ribosomal-RNAs that were sticking amino acids together at random, and the addition of transfer-RNAs and anchoring-RNAs gave origin to an apparatus that was still engaged in the synthesis of statistical proteins and was a long way from assembling specific proteins.

This is therefore the great difference between ancestral and modern genetic code: the ancestral genetic code was ambiguous and could not direct the synthesis of specific proteins. The modern genetic code, on the other hand, was based on protein-synthetases and had to preceded by a previous non-ambiguous code based on RNA-intermediaries that here is referred to as *ancient* genetic code. This tells us that there have been two distinct evolutions in the history of the genetic code: (1) the evolution from the ambiguous ancestral code to the non-ambiguous ancient code, and (2) the evolution from ancient to modern genetic code.

The population of primitive systems that gave origin to the genetic code is collectively referred to as *common ancestor*, but the above arguments make us realize that there have been at least two distinct types of such ancestors: the *first common ancestor* is the population that evolved the ancient genetic code whereas the *last common ancestor* is the population that gave origin to the modern genetic code.

## 4.10 The Ancient Genetic Code

The *ancestral* genetic code was ambiguous because any codon could code for various amino acids and a sequence of codons was inevitably translated into a set of statistical proteins. The evolution of this code was essentially a process that gradually reduced the *ambiguity* of the codons, i.e., the number of amino acids that they coded for, and came to an end when any codon was allowed to code for one and only one amino acid. This gave origin to the *ancient genetic code*, but how did it happen? What was the evolutionary advantage in reducing codon ambiguity?

The primitive systems were based on statistical proteins, and the ancestral genetic code was fully capable of directing the synthesis of these proteins, so it may seem that there was no need to reduce its ambiguity. Things, however, were different in that particular subgroup of statistical proteins that were associated with ribosomes, so let us take a closer look at them.

Ribosomes can be reconstructed by self-assembly from their isolated molecules, and it has been possible to discover the contribution of individual ribosomal proteins by studying what happens when ribosomes are reassembled without anyone of them in turn. These experiments have shown that the ribosomal proteins fall into three major groups: some are required for the assembly of the ribosome, others are necessary for its functions, and those of the third group have a stimulating effect but are fundamentally disposable (Kurland 1970; Fox 2010). Ribosomal proteins, in short, belong to three distinct families, and this suggests that the ancestors of these families were the result of three processes of diversification where some anchoring-RNAs started evolving into three families of ribosomal messenger-RNAs.

These messenger-RNAs were transmitted to new generations by copying, but their translation could reproduce the three ribosomal families of their progenitors only if the ambiguity of the genetic code was low enough to keep the differences between the translated proteins within the variability range of each family. Lowering the level of code ambiguity, in other words, had a great evolutionary value because it was allowing the reappearance of three distinct families of statistical ribosomal proteins in every new generation.

These three families, furthermore, had the potential to diversify into subfamilies, but again they could do so only if the ambiguity of the ancestral genetic code was further reduced in order to ensure that the differences between subfamilies would reappear in all future generations. We realize in this way that there was a real evolutionary advantage in reducing the ambiguity of the ancestral genetic code: the lower this ambiguity, the higher was the number of differences between the ribosomal proteins and the greater their ability to promote self-assembly and functional efficiency in the newly formed ribosomes.

We also realize that there was a distinct evolutionary advantage in promoting an overall increase in the number of ribosomal proteins. The reason comes from a general principle in engineering that Burks (1970) expressed in this way: *"there exists a direct correlation between the size of an automaton – as measured roughly by number of components – and the accuracy of its function"*. In our case, this principle means that increasing the number of ribosomal proteins was making the ribosomes more heavy, more resistant to thermal noise and therefore more accurate in protein synthesis.

The evolution of the genetic code, in other words, was *ribosome-oriented*: it consisted in a step-by-step reduction of code ambiguity that promoted a step-by-step diversification of the ribosomal-messenger-RNAs, and allowed them to direct the synthesis of an increasing number of different ribosomal proteins.

But do we have any evidence of this? Can we really say that the evolution of the genetic code went hand in hand with the evolution of the ribosomal proteins? This is equivalent to saying that the ribosomal proteins were the first specific proteins on Earth, and the evidence does appear to support this conclusion.

In all modern ribosomes there are a few proteins that are found in only one of the three primary kingdoms plus a large set of ribosomal proteins that are virtually universal (Woese 2002; Fox 2010). This means that the last common ancestor had not only a universal genetic code, but also a set of nearly universal ribosomal proteins. Which in turn means that the first common ancestor evolved not only the rules of the ancient genetic code but also most ribosomal proteins. The phylogenetic trees that are reconstructed from molecular data, on the other hand, do not reveal the existence of older proteins, and this strongly suggests that the ribosomal proteins were indeed the first *specific* proteins that appeared on the primitive Earth.

## 4.11 The Modern Genetic Code

The modern apparatus of protein synthesis came into being with the appearance of two great protein families: the *ribosomal-proteins* and the *synthetase-proteins* (aminoacyl-tRNA-synthetases). Paul Schimmel underlined that these synthetases are "*among the oldest proteins on the planet*" (Schimmel et al. 1993), and it is possible therefore that they appeared shortly after the ribosomal proteins. Since they are *specific* proteins, on the other hand, they could be produced only *after* the origin of the ancient genetic code, and before that event there had to be other intermediaries between amino acids and transfer-RNAs. The modern protein-synthetases, in short, replaced previous molecules (for example RNA-synthetases) and in so doing they could have modified the rules of the ancient genetic code. But did this actually happen?

In those early days most protein components of the ancestral systems were statistical proteins, and new coding rules would have merely reshuffled them. The changes would have been significant only in proteins with specific sequences and this means that only ribosomal proteins would have been affected. Changes in ribosomal proteins, on the other hand, were not necessarily a negative outcome. Ribosomal proteins had changed before and could well undergo other changes. It is true that there are universal features in them, but it is also true that they coexist with individual features that vary from species to species (in eukaryotes, for example, the ribosomal proteins are larger and more numerous than in prokaryotes, but this does not prevent them from performing equivalent universal functions).

As a matter of fact, the replacement of the old RNA-synthetases with protein-synthetases could well have *improved* the performance of the ribosomal proteins, and in this case there would have been an evolutionary advantage in replacing the ancient genetic code with the modern one. We also have another important argument in favour of this conclusion.

The experimental data suggest that the modern genetic code has been optimized for minimizing the impact of translation errors (Woese 1965; Woese et al. 1966), and various statistical studies (Haig and Hurst 1991; Freeland and Hurst 1998), have shown that the modern code performs better than virtually all its countless potential alternatives. These data leave little doubt that the modern genetic code went through a process of optimization, but it seems unlikely that its performance was tested and optimized for the synthesis of countless proteins. A much more realistic scenario is the idea that the genetic code was optimized for the synthesis of the 50 or so ribosomal proteins that existed at the time, and *once it was optimized for them it was automatically optimized for the synthesis of most other proteins.*

We realize in this way that the introduction of modern protein-synthetases in the apparatus of protein synthesis not only replaced the ancient genetic code with the modern one, but provided the means for optimizing its performance, a process that gave origin to a genetic code of extreme virtuosity and the accuracy of protein synthesis became so high as to be virtually error-free.

The ancient genetic code, in conclusion, evolved in parallel with the ribosomal proteins whereas the modern genetic code evolved hand in hand with the synthetase-proteins.

## 4.12 Steps Towards the Cell

The *ancient* genetic code allowed the ancestral systems to synthesize specific ribosomal proteins and obtain ribosomes that could be reassembled from the same components in every new generation. This ancient code, in turn, underwent a further evolution with the development of a second family of specific proteins, the aminoacyl-tRNA-synthetases, that optimized its performance in protein synthesis and gave origin to the *modern* genetic code. The last common ancestor is the population of primitive systems that evolved the modern genetic code, but were still a long way from a modern cellular organization. There have been at least three other major events before the origin of the first cells, and here is a brief summary of them.

1. *The switch to genetic code conservation*
   There are, in principle, 61 different transfer-RNAs, one for each codon, but the *wobble* mechanism, as we have seen, reduces their number to about 40 in most cells, and 40 different transfer-RNAs can carry 40 different amino acids. This means that the genetic code could have evolved all the way up to 40 amino acids, but it is an historical fact that its evolution stopped at the canonical set of 20. It is also an historical fact that the modern genetic code has been strongly conserved ever since, which means that the initial period of code evolution was followed by an immensely long period of code conservation. What is not usually underlined is that this conservation is necessarily an *active* process because the genes of the genetic code are constantly subject, like all other genes, to mutation and neutral drift and are in a continuous state of flux. The fact that they have been highly conserved means that there is a biological mechanism that *actively and continuously* restores their original structure. The conservation of the genetic code, in other words, is not the passive result of processes that somehow 'freeze' it. It can only be the result of an active biological mechanism that is continuously at work, a mechanism that has been referred to as *codepoiesis* (Barbieri 2012).

2. *The switch to protein metabolism*
   The ancestral systems *could* have continued synthesizing statistical proteins even when they acquired a modern genetic code, because a random sequence of codons is inevitably translated into a random sequence of amino acids whatever is the code. But they did not. It is an historical fact that other messenger-RNAs evolved and many other families of specific proteins came into being. The result was that protein enzymes gradually replaced the RNA enzymes and metabolism became increasingly a protein-based process. Biological specificity was extended to most proteins and this amounted to another major transition in the history of life: it transformed the old world of statistical proteins into the new world of specific proteins that gave origin to life as we know it.

3. *The development of new codes*
   The cells receive information from the external world by means of signal-transduction mechanisms that are based on organic codes, and the appearance of these codes was therefore a prerequisite for their origin. More precisely, as we will see in the next chapter, it was the appearance of three distinct signal-transduction codes that gave origin to the three cell domains, or primary kingdoms, that today are known as *Archaea*, *Bacteria* and *Eucarya*.

After the origin of the modern genetic code, in conclusion, there have been at least three other major transformations in the ancestral systems that prepared the ground for the origin of the first cells. But what is, precisely, a living cell?

## 4.13 The Ribotype Theory of the Cell

According to modern biology, the cell is a duality of *genotype* and *phenotype*. This model was proposed by Wilhelm Johannsen (1909) but was accepted only in the 1950s, when molecular biology made it clear that genes carry *linear information* whereas proteins function in virtue of their *three-dimensional structures*. The two components of the genotype-phenotype duality represent two different biological functions (heredity and metabolism), and two different physical quantities (information and energy), and describe the cell as a biological computer made of organic software and organic hardware. This is the simplest way of defining a living system, and has become the foundational model of modern biology, the scheme that transformed the *energy-based* biology of the nineteenth century into the *information-based* biology of the twentieth.

In the 1950s and 1960s, however, the study of protein synthesis revealed that genes and proteins are not formed spontaneously in the cell but are manufactured by a system of molecular machines based on RNAs. In the 1980s, the components of this manufacturing system were called *ribosoids* and their category was given the collective name of *ribotype* (Barbieri 1981, 1985). The cell was described in this way as a system made of genes, proteins and ribosoids, i.e., as a trinity of *genotype, phenotype and ribotype*.

The category of the ribotype may appear redundant, at first, because the ribosoids are usually regarded as components of either the genotype or the phenotype, but there is a good reason for keeping it distinct. The reason is that we can speak of genotype and phenotype only when a translation apparatus is present, because it is this apparatus that *defines* the genes and *synthesizes* the proteins. Genotype and phenotype, in other words, cannot exist without a translation apparatus based on RNAs. The consequence of this fact is that the ribotype is distinct from genotype and phenotype because it has both a logical and an evolutionary priority over them.

The RNAs and the proteins that appeared spontaneously on the primitive Earth produced a wide variety of ribosoids, some of which were synthesizing ribosoids whereas others were *ribogenes* and others were *riboproteins* (or ribozymes). The systems produced by the combination of all these molecules, therefore, had a ribotype, a ribogenotype and a ribophenotype. Eventually, evolution replaced the ribogenes with genes and the riboproteins with proteins but the ribosoids of the ribotype have never been replaced. This shows not only that the ribotype is a distinct category of the cell but also that it is a category without which the cell simply cannot exist.

It is an experimental fact, at any rate, that every cell contains a system of RNAs and ribonucleoproteins that makes proteins according to the rules of the genetic code, and that system can be described therefore as a *code-and-template-dependent-protein-maker*, i.e., as a *codemaker*. That is the third party that makes of every living cell a trinity of genotype, phenotype and ribotype. The genotype is the seat of heredity, the phenotype is the seat of metabolism and the ribotype is the codemaker of the cell, the seat of the genetic code.

# Chapter 5
# The First Three Billion Years

The *Cambrian* is the geological period whose rocks contain the first fossils that are visible to the naked eye, and the age that came before has been traditionally referred to as the *Precambrian*. At a visual inspection, the rocks of the Precambrian look totally devoid of traces of past life, but under the microscope some of them have revealed the presence of fossilized microorganisms (Schopf 1993). The Precambrian is therefore the geological age that came before the origin of visible multicellular creatures, and the dating of its rocks by radioactive methods has revealed that it has been an immensely long period.

The Earth is 4.6 billion years old and the first multicellular creatures appeared around 0.6 billion years ago, which means that the Precambrian went on for 4 billion years, almost 87 % of the entire history of the Earth. Fossilized Precambrian microorganisms, on the other hand, have not been found before 3.6 billion years ago, and this divides the Precambrian into two great eons: 1 billion years at the beginning, where there is no trace of biological systems, followed by 3,000 million years during which our planet was inhabited exclusively by microorganisms. That was *the Age of the cells*, an unimaginably vast expanse of time in which life was totally confined to the cellular level.

The data that allow us to reconstruct the history of life come primarily from paleontology and molecular biology. Paleontology uncovers the fossilized remains of past organisms whereas molecular biology allows us to reconstruct the phylogenetic relationships that exist between present and past organisms. The discovery of the genetic code, in particular, had direct phylogenetic implications because its presence in all living creatures means that that code appeared at the very dawn of life, in a population of primitive systems that has become known as the *common ancestor*.

The role of other organic codes in Precambrian evolution has largely been ignored, so far, and this chapter will try to fill that gap. More precisely, it will show that the presence of organic codes throughout the Precambrian gives us a new reconstruction and an entirely new understanding of the history of life.

© Springer International Publishing Switzerland 2015

M. Barbieri, *Code Biology*, DOI 10.1007/978-3-319-14535-8_5

## 5.1 The 'Stony Carpets'

In the early 1900s, Charles Walcott discovered a new type of rocks in North America. He found mounds, domes and pillars of various dimensions and shapes which had unique characteristics: they were made of wafer-thin layers, and were definitely Precambrian (Schopf 1999). The layers formed flat or curly mats, sometimes circular ones, but in all cases their pattern had, like in tree-rings, that irregularity of execution which is often a sign of biological activity. Walcott suggested just that: as whole islands have been built by corals, those domes and pillars had been left behind by Precambrian microorganisms. The cells had gone, but the minerals deposited on them had solidified and had become the stony replicas of the colonies. They were the empty fossilized cities of Precambrian empires.

A few enthusiasts supported Walcott's interpretation, and Kalkowsky (1908) gave to these formations their present name of *stromatolites* (stony carpets). Most geologists however were skeptical and diagnosed a non-biological origin for them. The search for Precambrian life went on for many more years amid failures, false alarms and general disbelief.

The breakthrough came in 1954, when Stanley Tyler and Elso Barghoorn examined under the microscope thin slices of rocks from the Gunflint deposits, near lake Ontario, and found a jungle of 2-billion year old microfossils (Tyler and Barghoorn 1954; Barghoorn and Tyler 1965). In this case, the cells had fallen into a solution of colloidal silica, or had been embedded in a limestone that was later replaced by silica, and in time the embedding matrix had crystallized into a very hard rock called chert. In a few cases, the most delicate details had been preserved, to the point that some microfossils could be given the status of distinct species. The discovery launched a world-wide hunt, and a few other fossilized gold-mines were found.

Then reports came that in some inhospitable places (for example at Shark Bay in Australia) there are stromatolites which still have living inhabitants, and whose microorganisms are strikingly similar to some of the types which had been buried 2 or 3 billion years earlier.

The picture of the Precambrian which has emerged from these discoveries is still fragmentary, but a few valuable conclusions can be drawn. The oldest terrestrial rocks come from the Isua mountains, in Greenland; they are 3.8 billion years old, and do not appear to contain fossilized cells. The oldest microfossils, so far, have been discovered in the Warrawoona formation, in Western Australia, and are approximately 3.4 billion years old (Knoll 2003).

These data do not come anywhere near a satisfactory dating of the origin of life, but they are all that we have at the moment. While keeping an open mind, therefore, the appearance of the first cells can be set somewhere in between the above figures, around 3.6 billion years ago. What is much more certain is that in the 3 billion years that followed life evolved exclusively at the cellular level.

## 5.2 The Iron Bands

After the excitement of the first discoveries, scientists had to face the problem of making a realistic assessment of Precambrian life. Did the microorganisms fill the oceans to capacity or were they condemned to a meager existence, like their descendants in Shark Bay? Was the Precambrian a Garden of Eden or a vast desert with only a few veins of life?

In the great majority of Precambrian rocks there are no signs of life, this is a fact. It is possible that these signs were erased later, when the rocks were greatly transformed by heat and pressure, but it is also possible that most rocks had never contained living creatures, and therefore that life had been extremely scarce in Precambrian times.

Luckily, there is a way of settling this problem. All over the world there are rocks from the middle Precambrian which consists of alternating layers of iron-rich and iron-poor silica (*Banded Iron Formations*), and these represent a planetary transition between two distinct phases in the geology of iron. In the absence of oxygen, iron is deposited in ferrous salts which are relatively soluble in water, while in aerobic conditions it produces salts of ferric iron which are insoluble. The sediments of the Iron Bands indicate therefore that a world-wide transition from ferrous to ferric salts was taking place, in other words that oxygen was accumulating in the atmosphere (Knoll 2003).

The process lasted nearly a thousand million years, roughly from 2.6 to 1.6 billion years ago, with a peak between 2.3 and 1.9 billion years ago which correspond to the bulk of the Iron Bands. The most important point, for us, is that the oxygen which changed the Earth's atmosphere could only have been released by microorganisms. It had to be then, as it is today, the waste-product of *photosynthesis* because there is no other natural phenomenon that could have caused a release of that magnitude (Cloud 1965).

The fossil remains of photosynthetic microorganisms, and in particular of *bluebacteria* (also known as *cyanobacteria* or *blue-green-algae*), have been found in Precambrian rocks, and there is no doubt therefore that they existed at that time, but we can conclude more than that. The transformation of the oceans and of the atmosphere of the planet was an operation of immense proportions, and could not have taken place if life was barely surviving is a few isolated areas.

This tells us that in the middle Precambrian the oceans were swarming with microorganisms, and in particular with bluebacteria, which are the most numerous forms in the Precambrian record and also the most versatile photosynthetizers. More precisely, we can say that the eon between 2.6 and 1.6 billion years ago was the age of the oxygen revolution on Earth. We can call it *Cyanozoic*, the Age of the bluebacteria.

## 5.3 The Age of the Protista

The greatest divide of the living world is not between plants and animals, as it has been thought for centuries, but between cells without a nucleus (*prokaryotes*) and nucleated cells (*eukaryotes*). On average, prokaryotes are definitely smaller than eukaryotes, and in general a typical prokaryote is even smaller than the nucleus of a typical eukaryote. Size, however, is only one of the many parameters that divide the two types of cells.

Prokaryotes, or *bacteria*, have a single DNA molecule arranged in a circle, a single cytoplasmic compartment where all biochemical reactions take place in solution, and the form of the cell is due either to an external wall that surrounds the plasma membrane or to a rigid plasma membrane.

Eukaryotes have a number of DNA molecules arranged in fibers which are repeatedly coiled and folded in highly organized chromosomes, and a composite cytoplasm which is divided into compartments and houses a variety of organelles (mitochondria, chloroplasts, lysosomes, endoplasmic reticulum, etc.); the form of the cell is not due to a surrounding wall but to an internal cytoskeleton which is made of three types of filaments (microtubules, microfilaments and intermediate filaments).

Virtually all eukaryotes need oxygen to live, and even the very few exceptions to this rule seem to be descendants of oxygen-dependent creatures that later adapted to anaerobic life (Schopf 1999). Among the prokaryotes, instead, we find an extremely wide variety of responses to oxygen. Some bacteria cannot tolerate it (*obligate anaerobes*), others can survive without it (*facultative anaerobes*) and a third class depend upon it absolutely, like most eukaryotes do (*obligate aerobes*).

These characteristics suggest that some prokaryotes evolved when there was no oxygen on Earth and others when the concentration of oxygen was changing, while true eukaryotes appeared only when the level of oxygen became stable and relatively high.

This conclusion is supported by the fact that from the end of the Cyanozoic to the Cambrian explosion of life – roughly from 1.6 to 0.6 billion years ago – the average dimensions of the microfossils become steadily greater than in all previous periods. Most Cyanozoic microfossils are less than 5 μm, whereas in the last billion years of the Precambrian, many fossilized cells are between 5 and 20 μm, and a significant percentage of them exceed 100 μm (Schopf 1978).

The increase in cell size that took place from the end of the oxygen revolution onwards, has a fairly unambiguous meaning: it signals the appearance of the first *modern* eukaryotic microorganisms, the *protista*. The fossil record, in other words, tells us that the last billion years of the Precambrian was characterized by the planetary expansion of these microorganisms and can therefore be referred to as the *Age of the protista*, a period that is traditionally known as *Proterozoic*.

The other great event in this period was the appearance, towards the end of it, of the first multicellular creatures. We cannot put a precise date on this event, but the fossil record shows that there was a substantial gap (at least 500 million years) between the first protista and the first multicellular organisms (Knoll 2003). The origin of modern eukaryotes and the origin of multicellular life, in other words, were definitely two separate historical events.

We can divide therefore the history of the Earth into five great chapters: four eons of about 1 billion years each (*Prezoic, Archeozoic, Cyanozoic* and *Proterozoic*), followed by a last eon of roughly 0.6 billion years that has been called *Phanerozoic*, the *Age of manifest life*.

## 5.4 Three Primary Kingdoms

Prokaryotes and eukaryotes have very different structures and lifestyles. Prokaryotes, or *bacteria*, are almost exclusively free-living single cells, and can inhabit virtually any ecological niche, with or without light, with or without oxygen, with or without organic material. Most of them are capable of synthesizing all organic components from inorganic molecules, can rapidly adapt to environmental changes, and exchange genes in a horizontal way, i.e., between individuals of the same generation. Bacteria are apparently capable of avoiding extinction, in the sense that some of them appear strikingly similar to those found in fossils that are more than 3 billion years old.

Eukaryotes, by contrast, are mostly oxygen-dependent organisms (many are obligate aerobes), and, in addition to monocellular forms (*protista*), have generated all three kingdoms of multicellular life (plants, fungi and animals). They invented new mechanisms of cell division (mitosis and meiosis), new types of movement, meiotic sexuality and finally embryonic development, a process that is potentially capable of generating countless different three-dimensional forms. In general, eukaryotes can adapt to the environment with highly sophisticated changes, but the price of their versatility is a very high level of extinction in the sense that the average lifetime of animal species is only 4 million years (Gould 2002). Prokaryotes and eukaryotes, in short, are two radically distinct forms of life, and we need therefore to understand the origin of the deep dichotomy that exist between them.

Schimper (1883) suggested that chloroplasts had once been free-living bacteria that happened to be incorporated, by a kind of symbiosis, into other cells, and from 1905 to 1930 this hypothesis was proposed to explain also the origin of mitochondria by Mereschowsky (1910), by Portier (1918) and by Wallin (1927).

The *symbiosis hypothesis* was largely ignored at first, but in 1970 it was forcefully re-proposed by Lynn Margulis, and within a few years it received the support of an astonishing number of experimental facts. It was found that mitochondria and chloroplasts are still carrying fragments of their ancient DNA, and have 70s ribosomes which are typical of bacteria, all of which leaves little doubt about their origin.

Today it is universally acknowledged that mitochondria and chloroplasts were acquired by symbiosis in the course of Precambrian evolution, but that tells us nothing about the cells that engulfed them, and on this issue biologists have been divided into two opposing camps. Some, like Lynn Margulis (1970), maintained that the cells which engulfed bacteria were themselves bacteria, whereas others insisted that they must have been eukaryotic cells or some of their precursors.

The solution came in 1977 from Carl Woese and George Fox with the discovery that the ribosomal-RNAs of all living creatures fall into three great groups: the rRNAs of two distinct types of prokaryotes that Woese and Fox (1977) called *archaebacteria* and *eubacteria*, and a third group containing the rRNAs derived from the ancestors of all modern eukaryotes that they called *urkaryotes*. Most importantly, they found that the RNA differences between archaebacteria, eubacteria and eukaryotes prove that each of these three groups has the same phylogenetic distance from the other two, which means that they had independent origins, and in particular that the ancestors of the eukaryotes *did not evolve from bacteria*.

Later on, Woese renamed the three groups and proposed that all cells belong to three distinct *primary kingdoms*, or *domains*, called *Archaea*, *Bacteria* and *Eucarya* (Woese 1987, 2000; Woese et al. 1990). The evidence of three distinct types of rRNAs implies that the progenitors of these kingdoms evolved independently from the population of primitive systems that has become known as the *common ancestor*, and this immediately gives us a problem: how did the common ancestor give origin to the first cells?

## 5.5 The First Cells

According to Eric Kandler (1994) and Carl Woese (2002), the evolution of the common ancestor was a *collective* process where the primitive systems were continuously exchanging genes between themselves and there were no effective limits to the process of horizontal gene transfer. This is why Kandler described the ancestral systems as "*metabolizing self-producing entities exhibiting most of basic properties of a cell, but unable to limit the frequent mutual exchange of genetic information*".

A similar concept was expressed by Woese with the idea that the common ancestor had an "*ephemeral cellular organization*", and the origin of the first cells was a process that created an "*increasingly permanent organization*" by putting a brake to unconstrained horizontal gene transfer. The transition from an ephemeral to a stable cellular organization was described by Woese as "the crossing of the *Darwinian threshold*", and he posited that the first cells originated by three distinct episodes of such crossing (Woese 2002).

It must be underlined, however, that horizontal gene transfer has continued to have an overwhelming impact in all cells, so much so that in *Archaea* and *Bacteria* it still accounts for up to 80 % of the genes (Dagan et al. 2008). The containment of horizontal gene transfer, in other words, could have played a role, but a reduction from nearly 100 % to 80 % does not seem enough to explain the origin of the first cells.

Luckily, another explanation does exist. An important clue comes from the fact that the combination of signal transduction with the genetic code allows a cell to regulate protein synthesis according to the signals from the environment and the cell acquires in this way a *context-dependent* behaviour (Jacob and Monod 1961). A signal transduction code was therefore of paramount importance to the ancestral systems, which makes it very likely that they made various independent attempts to develop it.

It is an experimental fact, at any rate, that *Archaea*, *Bacteria* and *Eucarya* have three different types of membranes and three distinct signalling systems, and this suggests that the three cell domains came into being by the combination of the universal genetic code with three distinct signal-transduction codes. This amounts to saying that a modern cell design requires at least two organic codes: a genetic code for protein synthesis and a signal transduction code for a context-dependent behaviour. The three primary kingdoms of life, in other words, were the result of three independent attempts performed by the descendants of the common ancestor to evolve a signal transduction code.

We realize in this way that the genetic code was instrumental to the origin of life and that the signal transduction codes were instrumental to the origin of the first cells. In these two cases, in other words, there has been a deep relationship between organic codes and the great events of macroevolution.

## 5.6 Two Evolutionary Strategies

In order to reconstruct the evolution of the first cells, let us start from the fact that bacteria appeared very early on our planet and some of them (in particular the *cyanobacteria*) have remained substantially the same ever since (Schopf 1999; Knoll 2003). The oldest known bacteria, in other words, were already similar to their modern descendants, and this allows us to draw at least three conclusions about the first cells.

1. The modern bacterial genome is a single molecule where all genes are arranged one after the other without interruptions and all of them carry information that is actually expressed. Such an organization is surely very efficient, but precisely for this reason it could not have appeared at the very beginning. A genome made of various DNA pieces where only a few were carrying useful information is definitely more primitive, and it is likely therefore that the first cells did have such genomes.
2. Bacterial protein synthesis is based on *unstable*, or *short-lived*, messengers that allow the cells to adapt very rapidly to changing environmental conditions Again, such a fast-reacting system cannot be primitive because unstable messengers require advanced forms of regulation and must have been the result of considerable evolution. It is likely therefore that the first cells were using more stable messengers because only these molecules are compatible with simple forms of regulation.

3. In bacteria, the transcription of the genes is immediately followed by translation, to the extent that in most cases protein synthesis starts on primary transcripts that are still attached to DNA. The result is that there is no time for a modification of the primary transcripts, and these are directly used as messenger RNAs. Such a fast sequence of precisely coordinated steps is hardly primitive and it is likely therefore that in the first cells transcription was physically detached from translation.

We realize in this way that the first cells did not have the typical features of modern bacteria, but much more primitive characteristics. What is most important is that because of these characteristics they had two evolutionary strategies in front of them, one based on increasing simplification, or streamlining, and one based on increasing complexity.

The cells that adopted a streamlining strategy got rid of all unnecessary components, arranged all genes on a single chromosome, evolved short-lived messengers, and abolished all steps between transcription and translation, thus acquiring the typical features of *modern bacteria*. They are divided into two great domains, *archaebacteria* and *eubacteria*, because they are adapted to two very different types of environments. The eubacteria are adapted to 'normal' environmental conditions, whereas archaebacteria are adapted to 'extreme' conditions and for this reason are also known as *extremophiles. Thermophiles*, for example, grow at temperatures between 80 and 120 °C, especially in oceanic and terrestrial hydothermal vents. *Psycrophiles* live in extremely cold environments, between 0 and 4 °C, and even stop reproducing when temperatures rise above 12 °C. *Halophiles* (salt-loving) grow in highly salty niches, such as salt-evaporation basins. As for pH, there are two different types of extremophiles: *basophiles* prosper in habitats, such as soda lakes, whose pH is greater than 9, while *acidophiles* colonize areas with pH between 1 and 5, like sulphur vents and rot undergrounds.

Bacteria, in short, have the ability to colonize virtually all environments on Earth and have had that ability since their early appearance on our planet. Not all ancestral cells, however, became bacteria. Some of them did not engage in streamlining, and followed the opposite strategy of increasing complexity. They were the ancestors of the eukaryotes, the cells that Woese and Fox (1977) called *urkaryotes*.

## 5.7 The RNA Codes

All DNA genes are copied into RNA molecules that are called *primary transcripts*, and all proteins are made from RNA molecules that are called *messenger RNAs*. In bacteria, the primary transcripts are directly used as messenger RNAs, but in most eukaryotes things are much more complex. The primary transcripts are first cut into pieces and then some of them (called *introns*) are removed and the remaining pieces (called *exons*) are joined together to form the messenger RNAs. These cutting-and-sealing operations are collectively known as *splicing*, and it has been shown that there are significant parallels between splicing and protein synthesis: (a) the splicing operators, known as *spliceosomes*, are huge molecular machines like ribosomes, (b) splicing employs small molecules, called *small nuclear RNAs* (*snRNA*), that are comparable to transfer-RNAs, and (c) both are processes that assemble molecules: splicing assembles messenger-RNAs from exons whereas protein synthesis assembles proteins from amino acids. The crucial point is that the choice of the beginning of an intron is completely independent from the choice of its end, which means that the snRNAs are real adaptors because they perform two independent recognition processes at each step of the reaction. In splicing, in other words, we find all the three essential components of a code: (1) two independents worlds of objects (primary transcripts and messenger RNAs), (2) a potentially unlimited number of arbitrary connections produced by adaptors, and (3) a set of coding rules (a selection of the adaptors) that ensures the specificity of the correspondence.

From an evolutionary point of view, what is most important is the fact that splicing requires a separation *in time and space* between transcription and translation, and this was a precondition for the development of a physical barrier between them, a barrier that eventually took the form of the nuclear membrane. The origin of this membrane – and with it the origin of the nucleus – is traditionally regarded as the origin of *modern eukaryotes*, and the splicing codes were instrumental in bringing about that major transition in evolution.

Splicing, however, is not the sole mechanism that acts on messenger-RNAs. Another operation is *RNA editing*, the process that alters the RNA sequences by replacing one or more nucleotides in their codons, thus altering the amino acid sequence of the encoded protein (Brennicke et al. 1999; Bass 2002). A classical example of editing is the modification of Apolipoprotein-B (ApoB), the protein that carries lipids in the tissues of the body. In the liver, this protein is synthesized in the normal form (ApoB 100) but in the small intestine it is produced in a truncated form (ApoB 48) because an operation of RNA-editing transforms the codon for glutamine (CAA) into a stop codon (UAA) that brings protein synthesis to an end. Both proteins are active but perform different functions in the transport of lipids, which shows that RNA editing can have regulatory roles.

The operations of RNA editing require the intervention of supramolecular particles called *editosomes* that are somehow comparable to *spliceosomes* and *ribosomes* (Madison-Antenucci et al. 2002; Panigrahi et al. 2003), and this suggests that editosomes too operate on the basis of coding rules. So far, however, the search for molecular adaptors in RNA-editing has not been carried out and the existence of *editing codes* remains at the moment a theoretical possibility.

In addition to messenger-RNAs and transfer-RNAs, a whole population of small RNA molecules have been found in living cells and are known as *micro RNAs*, *small interfering RNAs*, *small nuclear RNAs* and *small nucleolar RNAs* (Ambros and Chen 2007; Amaral et al. 2008; Bartel 2009). These particles intervene in virtually all aspects of gene regulation (Mattick 2004) and this makes it likely that their operations are based on coding rules, but again we do not have, at the moment, any direct evidence on such rules.

The family of the RNA codes, in conclusion, is an extremely heterogeneous component of the living world. The genetic code and the splicing codes are its most important members but around them there is a whole galaxy of RNA molecules and operations that still represent a largely unexplored field of research.

## 5.8 The Fluid Genome

The *Chromosome Theory of Heredity*, proposed by Sutton (1903) and Boveri (1904), stated that genes are carried on chromosomes, but did not say anything about their positions. It was Thomas Hunt Morgan (1915) who picked up that problem. In order to build maps of the chromosomes he needed to establish gene positions, and chose the simplest hypothesis: the idea that genes are arranged in linear sequences and do not move around but occupy *fixed* positions on chromosomes, like passengers which are glued to their seats in train carriages. This is the classical picture of genes that behave in a *Mendelian way*.

The very success of Morgan's maps gave an implicit endorsement to the underlying hypothesis, and so the *fixity of gene position*, i.e., the *Mendelian behaviour of genes*, was accepted as an article of faith and became the foundational hypothesis of genetics.

In the 1940s, however, while studying the genes that affect the colour of kernels in corn, Barbara McClintock noticed that some white kernels were carrying many scattered plum-coloured spots. In the cells responsible for those spots, in other words, some mutations had taken place which turned the colour from white to plum. The unexpected was not the mutations *per se*, but their astonishing frequency. The natural rate of mutation is about *one in a million* for any gene, and yet McClintock was observing the same mutation taking place simultaneously in thousands of cells of the same plant in just one generation.

The only solution she could come up with (McClintock 1951, 1956) was the hypothesis that the mutations were propagated by genes physically *jumping* from one chromosome to another. There was enough for burning her on the stake for heresy, and in a metaphorical sense that is what happened, as she herself recollected after receiving the Nobel prize in 1983. The Nobel prize was of course a recognition that she had been right all along, a fact that became universally accepted only when mobile genes, also known as *transposable elements* or *transposons*, were discovered not only in plants but in all other forms of life, from bacteria to animals.

And that was only the first of a series of discoveries which followed in quick succession and literally opened the gates of *non-Mendelian heredity*. Various other processes – such as *unequal crossing-over*, *DNA slippage* and *gene conversion* – proved that the genome is actually a dynamic superstructure where genes are in a continuous state of flux, a system that has become known as *fluid genome* (Dover and Flavell 1982). But what is there behind these processes? Are they based on chemistry alone or on coding rules?

So far we do not have an answer to these questions. What we know is that the transposition of genes from one place to another is carried out first by *transposase* enzymes that cut a piece of DNA at a target site and then by *ligase* enzymes that reinsert it at a different site. But a description is not a mechanism. It is as if we were describing a game of chess by recording the movements of the pieces but ignoring the rules of the game.

A similar situation exists in the case of many other processes that concern the use of genetic information. What is clear, today, is that the Mendelian behaviour of genes is only a crude approximation of the truth, and that the genome is not a *read-only-memory*, but a *read-and-write* memory (Shapiro 2011). So far, however, we have only described what goes on in the fluid genome, and the search for its coding rules is still a largely unexplored field of research.

## 5.9 Evolving the Cytoplasm

A process of biochemical evolution started fairly early on our planet and slowly transformed it into a niche where the first cells appeared together with countless organic compounds and nutrients. Even a large food supply, however, was destined to get depleted, and this created the conditions for the appearance of two very different survival strategies. Some cells adapted their metabolism to smaller and smaller molecules, and eventually learned to perform all metabolic reactions from inorganic matter. In this way they ceased to be *consumers*, and became *producers* of organic matter.

Other cells, however, continued to feed on organic materials and the increasing scarcity of small nutrients forced them to ingest increasingly bigger compounds. A potentially important source of food was the bodies of other cells, especially dead ones, and the ancestral consumers learned to develop structures that enabled them to ingest bigger and bigger pieces of organic matter. Such a property required a plasma membrane that could engulf a prey, and to that purpose the cells had to be able to change their shape *from within*, by using molecular structures that eventually evolved into *cytoskeletons*.

Once in existence, on the other hand, a primitive cytoskeleton was potentially capable of many other operations and that probably explains why it acquired a variety of different functions in the course of time. In addition to the ability to ingest a prey while standing still, for example, the cytoskeleton evolved the means to perform the movements that allow a cell to actively go searching for a prey rather than passively waiting for it. In parallel with external movements, furthermore, the cytoskeleton evolved a number of internal movements, such as those that take place in mitosis and are responsible for the physical displacement of the chromosomes to the opposite end of the mitotic spindle. The cytoskeleton also had a crucial role in the process by which two cells can fuse together in a single system that maintains the ability to divide itself, a step that was a precondition for the *origin of sex*, one of the greatest events of macroevolution that took place in the Precambrian.

Another outstanding Precambrian invention was the development of cell compartments that perform specialized functions with a great degree of autonomy, and at the same time work together in an integrated whole. This subdivision of the eukaryotic cell into compartments was accompanied by a parallel subdivision of the genome into largely autonomous clusters of genes, and set the stage for the future subdivision of the animal body into organs and for the differential expression of their genes.

There is a large consensus, today, that the evolution of the cytoskeleton and of the compartments were major transitions in the history of life, and yet there is something vitally important about them that is not usually appreciated. It is largely ignored, today, that their operations are based on organic codes despite the fact that they employ molecular intermediaries that have the typical characteristics of adaptors.

## 5.10 The Cytoskeleton Codes

A cytoskeleton is an internal frame of filaments that allows eukaryotic cells to perform a variety of dynamic processes such as phagocytosis, mitosis, meiosis, ameboid movement and change of their three-dimensional form. The actual cytoskeleton is the result of three different frames made of filaments (*microfilaments*, *microtubules* and *intermediate filaments*) each of which gives a specific contribution to the mobility of the cell and to its three-dimensional structure.

The driving force of the cytoskeleton is a special mechanism that biologists have called *dynamic instability*. The cytoskeletal filaments – especially microtubules and microfilaments – are in a state of continuous flux where monomers are added to one end and taken away at the other, so that the filament is growing or shortening according to which end is having the fastest run. But what is really surprising is that all this requires great amounts of energy, which means that the cell is investing energy not in building structures but *in making them unstable!*

In order to understand the logic of dynamic instability, we need to keep in mind that cytoskeletal filaments are unstable only when their ends are not attached to particular molecules that have the ability to anchor them. Every microtubule, for example, starts from an organizing center (the *centrosome*), and the extremity which is attached to this structure is perfectly stable, whereas the other extremity can grow longer or shorter, and becomes stable only when it encounters an anchoring molecule. If such an anchor is not found, the whole microtubule is rapidly dismantled and another is launched in another direction, thus allowing the cytoskeleton to explore all cytoplasm's space in a short time.

A classic example of this strategy is offered by mitosis. In this case it is imperative that microtubules become attached to *centromeres*, so that chromosomes can be transported to the opposite ends of the spindle, but centromers are extremely small and their distribution in space is virtually random. Looking for centromeres is literally like looking for a needle in a haystack, and yet the exploratory mechanism of dynamic instability always finds them, and always manages to find them in a brief span of time.

Now the logic is beginning to emerge. Dynamic instability is a mechanism that allows the cytoskeleton to build structures with an *exploratory strategy*, and the power of this strategy can be evaluated by considering how many different structures it can produce. The answer is astonishing: the number of different cellular forms that the cytoskeleton can create depends upon the combination of the anchoring molecules, and is therefore potentially *unlimited.* It is the anchoring molecules (that strangely enough biologists call *accessory proteins*) that determine the form that eukaryotic cells can have in space and the movements that they can perform. The best proof of this enormous versatility is the fact that the cytoskeleton was invented by unicellular eukaryotes, but later was exploited by metazoan to build completely new structures such as the axons of neurons, the myofibrils of muscles, the mobile mouths of macrophages, the tentacles of killer lymphocytes and countless other specializations.

Dynamic instability, in conclusion, is a means of creating endless cellular forms and movements with only one common structure and the choice of a few anchoring molecules. But this is possible only because there is no necessary relationship between the filaments of the cytoskeleton and the cellular structures that they are attached to. The anchoring molecules (or accessory proteins) that create a bridge between them are true adaptors that perform two independent recognition processes: microtubules on one side and various cellular structures on the other side. The result is based therefore on arbitrary rules, on true natural conventions that have been referred to as *cytoskeleton codes* (Barbieri 2003).

## 5.11 The Compartment Codes

The cell membrane of bacteria is like a molecular skin because it synthesizes its molecules *in situ*, just as a skin contains the cells that continually renew it. In eukaryotes, instead, the cell membrane is produced by a completely different mechanism. The membrane replacements are made in the interior of the cell in the form of vesicles that move towards the surface and here become incorporated into the existing membrane, while other vesicles detach themselves from the plasma membrane and move towards the interior to be recycled. In eukaryotes, in other words, the plasma membrane is the continuously changing result of two opposite flows of vesicles, and its integrity is due to the perpetual motion of these ascending and descending currents.

This mechanism may appear unnecessarily complex, especially when it is compared with the simplicity of the bacterial one, but this is only a first impression. Its true logic comes immediately to light as soon as we regard it not as an isolated case, but as an example of a wider class of phenomena. More precisely, as one of the various mechanisms that eukaryotic cells employ to build their *compartments.* The vesicles that are destined to the plasma membrane, in fact, are produced in the Golgi apparatus together with vesicles which have very different destinations. Some are delivered to lysosomes and others to secretory granules.

The Golgi apparatus is involved in the terminal modification of innumerable molecules which have diverse destinations, and if every molecule had to follow a specific path, the cell simply could not cope with the immensely intricate traffic that would have to be directed. The Golgi apparatus, instead, delivers to their destinations an astonishing number of molecules with only two types of vesicles: one for transporting proteins outside the cell, and the other to its interior. This requires only two destination signals for the vesicles, however large is the number of transported proteins. On top of that, the Golgi apparatus produces a third type of vesicles which do not carry any destination signal, and these are the vesicles that are programmed, *by default*, to reach the cell membrane. As we can see, the solution is extraordinarily efficient: with a single mechanism and only two types of signals, the cell carries an enormous amount of specific products to their destinations, and also manages to continually renew its membrane.

The Golgi apparatus, however, is a transit place for only a fraction of the proteins which are produced in eukaryotic cells. The synthesis of all of them begins in the soluble part of the cytoplasm (the *cytosol*), and during this first step they also receive a signal that specifies their geographic destination. The piece of the amino acid chain that emerges first from the ribosome machine – the so-called *peptide leader* – can contain a sequence that represents an *export signal to the endoplasmic reticulum.* If such a signal is present, the ribosome binds to the reticulum and delivers the protein into its *lumen.* If the peptide leader does not carry that signal, the synthesis continues on free ribosomes, and the resulting proteins are shed in the cytosol. Of these, however, only a fraction are destined to remain there, because a peptide chains can carry, in its interior, one or more signals which specify other destinations. More precisely, there are signals for protein export to the *nucleus*, to *mitochondria*, and to other cell compartments. Proteins, in conclusion, carry with them the signals of their geographic destination, and even the absence of such signals has a meaning, because it implies that the protein is destined to remain in the cytosol.

The crucial point is that there is *no necessary link* between protein signals and geographic destinations. The export-to-the-nucleus signals, for example, could have been used for other compartments, or could have been totally different, just as the names which are given to cities, to airports and to holiday resorts. The existence of eukaryotic compartments, in conclusion, is based on natural conventions, and to these rules of correspondence we can legitimately give the name of *compartment codes* (Barbieri 2003).

## 5.12 The Tree and the Web of Life

Darwin (1859) described evolution as a process that started "*from simple beginnings*" and gave origin to increasingly diverse "*most beautiful forms*", much like a tree that grows and divides into countless branches. At the basis of the tree there are the first "*primordial forms*" – the first cells – and out of them grows a trunk which splits again and again to create an ever expanding tree. Each branch represents a species and the branching points are where one species gives origin to two. Most branches come to a dead end, signifying extinction, but some go all the way up to the top and are today's living creatures. This is the *Tree of Life*, the graphic description of how every species is related to all the others that have appeared on Earth since the origin of life.

The reconstruction of the tree of life has been a holy grail for generations of naturalists, but then the unexpected happened and today the project has changed almost beyond recognition. The unexpected came in the 1990s as a result of the discovery that bacteria are routinely swapping genetic material with many other organisms in a process called *horizontal gene transfer* (Miller 1998). The pattern of a genealogical tree is realized when genes are passed down from one generation to the next, i.e., when descent is *vertical*. When genes are swapped *horizontally* between co-existing species, instead, they become part of many branches simultaneously and the resulting pattern is no longer a tree but a web (Doolittle 1999).

It turned out, furthermore, that horizontal gene transfer is by no means a secondary process. In bacteria it can account for as much as 80 % of the genes (Dagan et al. 2008), whereas in unicellular eukaryotes (protista) it is only slightly lower than that (Andersson 2005), and in animals and plants – where genes are transferred horizontally by viruses – it can make up for 40–50 % of the whole genome (Doolittle and Bapteste 2007).

The message that come from the experimental data is clear enough: for the first 3,000 million years of evolution there has been no proper tree of life but a vastly interconnected web. A tree-like pattern began to emerge only in eukaryotes, where the effects of vertical descent became increasingly evident over the levelling drive of horizontal gene transfer. The web of life, on the other hand, has always been, and continues to be, the majority pattern in the living world, and this makes us realize that there are three distinct forms of life on our planet.

The molecules responsible for horizontal gene transfer are nucleic acids (RNAs and DNAs) that are produced by copying and that survive by invading any system that is capable of copying them. They represents a form of life based on 'copying alone' and since this is the logic of the viruses we can give them the collective name of *viroidea*. Bacteria are cellular systems that are based on copying and coding, but have adopted a streamlining strategy that prevented them from developing new organic codes. They represent a second form of life, known as *prokarya*, that is based on 'copying and *limited* coding'. Other cells, the *eukarya*, have maintained the potential of the ancestral systems to explore the coding space and represent a third form of life based on 'copying and *unlimited* coding'.

Eukarya, Prokarya and Viroidea, in conclusion, represent distinct forms of life, and there have been two parallel drives in evolution: one that created a web-like pattern and one that generated the classical branching pattern of the tree of life. The global web of viruses and bacteria accounts for more than 90 % of the mass of organic matter that exists on Earth (Sonea 1988) and is by far the dominant form of life on our planet. The fact remains, however, that life managed to go beyond the level of viruses and bacteria and to give origin to increasingly complex forms that appeared together with new organic codes. Copying alone was not enough, and that again makes us realize the extraordinary importance that coding, and in particular *unlimited* coding, had for the evolution of life on our planet.

## 5.13 What Happened in the Precambrian?

The idea that the first cells were bacteria and that the first eukaryotes were aggregates of bacteria is still popular, today, but ever since the pioneering work of Woese and Fox (1977) the molecular phylogenetic data have increasingly supported the conclusion that three distinct types of cells evolved independently from the common ancestor, and in particular that the progenitors of the eukaryotes did *not* evolve from bacteria.

It must be underlined that this conclusion does in no way belittle the extraordinary importance of bacteria in the history of life. It was they that physically transformed the Earth into a blue oxygenated planet and created the *biosphere*, the spherical layer around the globe that provides the habitat for all forms of life. And it is the bacteria that still support all other creatures by continuously recreating the biosphere.

Bacteria, furthermore, are masterpieces of biological engineering. They are the best known example of systems capable of *autopoiesis*, i.e., systems that fabricate themselves and conserve their organization for millions of years. They are continuously evolving and yet fundamentally they hardly change. They are probably the nearest thing to immortality that exists in the living world, and it is precisely for this reason that bacteria could not be primitive. There had to be much less perfect creatures before them, and we have seen in fact that the common ancestor was made of supramolecular systems that evolved the rules of the genetic code and were engaged in *codepoiesis*, not in autopoiesis.

The bacteria were the crowning achievement of a strategy of extreme streamlining. All steps between transcription and translation were eliminated so that the primary transcripts of the genes could be directly used as messenger-RNAs without any intermediate processing. Their external form was determined either by a rigid wall around the cell membrane or by a rigid cell membrane, which avoided the need the produce an internal cytoskeleton, internal movements and internal compartments.

But not all descendants of the common ancestor became bacteria. Some of them continued exploring the coding space and evolved new organic codes such as splicing codes, cytoskeleton codes and compartment codes. Eventually, they acquired mitochondria and chloroplasts by symbiosis and gave origin to modern monocellular eukaryotes, the cells that are usually referred to as *protista*. And at this point the old bifurcation between codepoiesis and autopoiesis came back in a new form. Most protista stopped evolving new organic codes and became autopoietic systems that have remained substantially the same ever since. Only a few of them did not give up the ancestral strategy of codepoiesis, but the eukaryotic cell had already become the most complex system in the universe, and the only solution was exploring the new space of multicellularity.

That, in a nutshell, is what happened in the Precambrian. It was the bacteria who conquered the Earth and turned it into a living planet, but it was the ancestral strategy of codepoiesis that gave origin first to the common ancestor, then to the first cells and after that to the 3,000-million-year-long story that explored the space of complexity and finally culminated in the Cambrian explosion of multicellular life. It was new organic codes, in other words, that brought into being the great novelties of macroevolution.

# Chapter 6
# Evolving the Embryos

The most elegant and probably the most influential experiment in embryology was performed by Aristotle some 2,400 years ago on chick embryos. Aristotle removed the shell from eggs that had been incubated for different periods of time, and carefully described what he saw in successive stages of development: the white spot on the yolk that marks, at the beginning, the position where the embryo is going to appear; the tiny brown lump that begins pulsating at the third day; the protruding bulbs that gradually turn into eyes and brain vesicles; the network of red vessels that descend into the yolk and expand in it like the roots of a tree, and the various membranes that surround the growing chick (Aristotle [384–322 BC] 1965).

On the basis of these observations, Aristotle concluded that embryonic development is a step-by-step generation of new structures. It is not a single genesis but an *epigenesis*, a sequence of many geneses one after the other. An embryo, in other words, is a system that grows not only in size but also in number of parts and therefore in complexity.

Aristotle underlined that there are *two* distinct features in epigenesis. One is the increase in complexity, the other is that such increase goes in a very specific direction. A frog embryo, for example, can only become an adult frog and nothing else. Epigenesis, in other words, is the idea that embryonic development is not just an increase in complexity, but a *convergent* increase in complexity, and we need therefore to understand how it can be 'oriented' or 'channelled' towards a specific result.

To this end, Aristotle proposed a theory that has become the object of countless controversies ever since: the theory that in every living creature there is a *design*. Take a house, for example. When you build one, said Aristotle, you have no choice but to follow the laws of physics, those eternal rules that the old philosophers called *necessity* because nobody, not even the gods, can change them. They are the *efficient* causes of the phenomena and provide the physical mechanism in any natural process. In order to build a house, however, you need more than the laws of physics. You need a project, a design: this is the *final* cause of your enterprise, the goal towards which you move in full agreement with the laws of physics.

© Springer International Publishing Switzerland 2015

M. Barbieri, *Code Biology*, DOI 10.1007/978-3-319-14535-8_6

The great difference between building a house and building an embryo is that the design of a house comes from an external architect, and therefore *from without*, whereas the design of an embryo comes *from within* the system. Nature, in other words, does not build as we do. She is rational, because she works according to designs, but her plans are *immanent* in her creatures. They do not come from another world, as Plato said; they are an integral part of this world, just as matter is.

The contribution of Aristotle to embryology consists therefore in two great ideas: the *theory of epigenesis*, the doctrine that development is a convergent increase in complexity, and the *theory of design*, the idea that embryonic development is guided by a project that exists in the embryo from the beginning of conception.

Jacques Monod (1970), Max Delbrück (1971), Ernst Mayr (1982) and many others have pointed out that the immanent design of Aristotle is virtually equivalent to the genetic programme. The one important difference is that Aristotle's design was supposed to act from the future to the past, whereas the genetic programme works from the past to the future, in the same direction as the arrow of time.

It is an historical fact, however, that at the very beginning of the scientific revolution the Aristotelian theories of epigenesist and immanent design were banned from science, and the history of embryology has largely been the history of the discoveries that have reintroduced them in biology in a modern form.

## 6.1 The Rediscovery of Epigenesis

The experiment performed by Aristotle on chick embryos was repeated, roughly 2,000 years later, by Marcello Malpighi, in the 1660s, but this time with an important novelty. Malpighi observed the embryos under a *microscope*. He made detailed drawings of what he saw and found that many embryonic regions that look totally devoid of structures to the naked eye, contain in reality tiny blood vessels. This allowed him to conclude that Aristotle had been wrong. The structures of the embryos that seem to appear *ex novo* are in fact already present in microscopic form before they become visible to the naked eye (Malpighi 1673).

Malpighi became in this way one of the first proponents of the theory of *preformation*, the idea that embryonic development is merely the growth of an organism whose structures are already *preformed* in the fertilized germ. This idea spread very rapidly and was enthusiastically accepted by almost all naturalists of the seventeenth and eighteenth centuries. Swammerdam, Leeuwenhoek, Leibnitz, Réaumur, Spallanzani, Boerhaave, von Haller, Bonnet and many other great scientists declared themselves convinced preformationists.

Today we find it difficult to understand this popularity, but we should not forget that atoms and molecules were not known at the time, and the infinitely small was as legitimate as the infinitely large. The theory of preformation, furthermore, was avoiding the great mystery of the *mechanism* of epigenesis. Charles Bonnet, for example, could find no other rational answer and in 1764 made a passionate plea for preformation: "*If organized bodies are not preformed, then they must be formed every day, by virtue of the laws of a special mechanics. Now, I beg you to tell me, what mechanics will preside over the formation of a brain, a heart, a lung, and so many other organs?*" (Bonnet 1764).

Eventually, however, chemistry revealed the existence of atoms and molecules and it became clear that the reduction of biological structures to smaller structures cannot go on indefinitely. The official turning point came in 1828, when Karl Ernst von Baer published '*On the Development of Animals*', a monumental treatise of comparative embryology that put an end to any version of preformationism (von Baer 1828).

Von Baer showed that there is a very early stage in the development of all animal species where the entire embryo consists in a few sheets, or *germinal layers*, of organic matter that later became known as *ectoderm*, *mesoderm* and *endoderm*. After that stage, furthermore, the anatomical evidence clearly shows that the development of an embryo is not only a process of growth, but also a step-by-step emergence of new histological tissues, and a sequence of three-dimensional movements that deeply transform the shape of the developing creature.

With the advent of the cell theory, in the 1830s, embryonic growth was immediately accounted for by the process of cell division. A fertilized egg divides into 2 cells and then into 4, 8, 16 and so on. With 10 divisions the cell number is about a thousand, with 20 is a million, with 30 is a billion, with 40 is a thousand billion, and so forth. The difference between an adult body and a fertilized egg, however, is not a mere question of cell numbers. Fifty thousand billion eggs, whatever their arrangement in space, would never make a human being, and it is clear therefore that during development the cells must become *different* from the fertilized egg. Embryonic development is accompanied therefore by a hierarchy of differentiation processes which may result in more than 200 different types of cells (Alberts et al. 2007).

During development, furthermore, the form of an embryo undergoes many transformations before one can start recognizing the familiar features of the future adult organism. These changes are brought about by migrations, invaginations, tubulations and folding, and are collectively known as *morphogenesis*.

The experimental evidence, in short, allowed von Baer to reach a conclusion that is rightly regarded as the starting point of modern embryology: *embryonic development is a true epigenesis and consists in three fundamental processes – growth, differentiation and morphogenesis.*

## 6.2 Body Plans and Animal Phyla

Fossils have been known for many centuries, but in most cases they consist in a few bones that apparently tell us very little about past organisms, and this is why Paleontology has remained little more than a curiosity until the end of the eighteenth century. What transformed it into a science were the rules that George Cuvier, the founder of comparative anatomy, introduced between 1795 and 1828. The breakthrough came with Cuvier's first golden rule, the *Principle of the Correlation of Parts*. This principle tells us that it is possible to reconstruct a whole body from a small part of it because every part is correlated with the others, and in general it is compatible with only one anatomical architecture (Cuvier 1800–1805). Cuvier applied this principle to many cases and proved that the anatomy of entire organisms can indeed be reconstructed from a few bones. He was also able to demonstrate that the elephant-like creatures that had been found in Siberia belonged to a species which is different from all existing elephants, the extinct species of the mammoths.

The second great contribution of Cuvier was the introduction of a new biological category that he called *embranchement* and that today is known as *phylum* (Cuvier 1812, 1828). He made a comparative study of many animal species and discovered that their immense variety of body-structures can be reduced to a few basic anatomical designs. In the wings of birds and bats, for example, we find the same type of bones that exist in the arms of primates, in the flippers of seals and in the forelimbs of reptiles and amphibians. The size and the shape of each bone can vary enormously from one species to another, and yet we can clearly recognize which is which simply by its position and its relationships with the other bones. This is expressed by saying that the forelimbs of all vertebrates are *homologous*, because they have the same structural design, as if they had been obtained by modifying a common prototype-limb.

The wings of insects, on the other hand, do not have bones. They are *analogous* to the wings of birds because they are used for the same function of flying, but their structural plan is totally different. The key point is that a structural difference in a single key part means that the whole anatomical plan of the body is different. A *body plan*, in short, is a set of anatomical characters that describe the spatial organization of the body structures, and a *phylum* is a group of organisms that have the same body plan.

By a strange coincidence of history, the concept of phylum was also discovered by Karl Ernst von Baer, with a totally different method. Instead of adult animals, von Baer was studying the embryos of different classes of vertebrates, and discovered that there is a stage in development where the embryos of fishes, amphibians, reptiles, birds and mammals are practically indistinguishable (it seems that the discovery took place by chance when one evening von Baer forgot to label the bottles containing the embryos, and the next day could no longer recognize which class they belong to).

Von Baer concluded that all animals can be divided into major groups – today called *phyla* – which are characterized by the fact that the animals of one group have a stage of embryonic development in which all their embryos look similar. This stage of *maximum morphological similarity* is known today as the *phylotypic stage*, precisely because it is a defining characteristic of each phylum.

The link between the two discoveries is the fact that the phyla defined by Cuvier on the basis of the adult body plans, exactly correspond to the phyla defined by von Baer on the basis of the embryonic phylotypic stages, a convergence that has a clear biological meaning. It is due to the fact that the body plan is built in the course of embryonic development, and the phylotypic stage is the period in which all its defining features are most visible.

The fact that the body plan of an organism is already present in the embryo may suggest a similarity with the immanent design of Aristotle, but this is not the case because the body plan in built in stages whereas the design of Aristotle was supposed to be present since the moment of conception. What is similar to Aristotle's design, in fact, is the *genes* of the body plan, but they were not known at the time of von Baer.

## 6.3 A Cascade of Inductions

The microscopic description of the embryos in successive stages of development gives us a detailed picture of the changes that accompany the formation of new structures. A classical example is the sequence of events that occur in the development of the eye in vertebrates. Soon after the appearance of the neural tube, a bud is formed on its two sides, in the head region, and grows into a thin filament that moves outwards and become the precursor of the future *optic nerve*. The top of the filament forms a spherical expansion, the *optic vesicle*, and when this vesicle reaches the ectoderm a new set of transformations takes place. The optic vesicle changes its round shape into a concave one and becomes the *optic cup*, the precursor of the *retina*; simultaneously the overlying ectoderm bends towards the optic cup, and moves inside forming a ball of tissue that eventually detaches itself and becomes the *lens*. This sequence of events *suggests* that the optic vesicle is causing the transformation of the overlying ectoderm into a lens, but does not *prove* it. A description does not reveal the mechanism of a phenomenon and in order to find this out it was necessary to perform experiments.

This is precisely what Hans Spemann set out to do in the early 1900s. He invented a glass needle technique that allowed him to perform microsurgery operations, and made a systematic series of experiments on many developing organs. In the case of the eye, Spemann was able to remove the optic vesicle before it reaches the ectoderm and showed that in this case no eye is formed. Then he transplanted an optic vesicle under the ectoderm of the embryo's flank and found that a lens is formed there. This proved that the optic vesicle is the causal agent that *induces* the overlying ectoderm to become a lens (Spemann 1901).

Spemann then discovered that the lens, once formed, becomes a new inducing tissue and sends out signals that induce the overlying ectoderm to differentiate into a cornea. He proved, furthermore, that the optic vesicles itself had been the result of a previous induction. One of the very early tissues that appear in the embryo is the *dorsal cord*, and Spemann found that the signals emitted from this tiny mesodermal rod induce the overlying ectoderm to become neural tissue. More precisely it is a specific segment of the dorsal cord that induces the above segment of the neural tube to send out the filament that carries the optic vesicle.

Spemann proved in this way that most tissues of the body are formed by a *chain of inductions* where an induced tissue becomes in turn the inducer of another tissue. Such a chain, or cascade, on the other end, must have a beginning, and the discovery of that starting event turned out to be a milestone in embryology. The experiment in question was actually performed by Hilde Mangold, as part of her Ph.D thesis with Spemann, but it was rightly published under both names (Spemann and Mangold 1924).

The experiment was performed on salamanders at the gastrula stage, when the embryo is a hollow sphere of several thousand cells which starts a complex movement of invagination at a spot known as the '*upper lip of the blastopore*'. Hilde Mangold transplanted the upper lip from one embryo to the flank of another one, in the expectation of discovering which tissue is induced in the host embryo, but the result was a complete surprise. Three days after the transplant, an entire secondary embryo had formed on the flank of the host. The upper lip of the blastopore had set in motion not a single induction but a whole cascade of inductions. It was capable of organizing first a neural tube and then a fully formed body around it, and for that reason Spemann gave it the name of *organizer* (Hamburger 1988).

The presence of an organizer was later found in many other vertebrates (in birds and mammals, for example, it is the structure known as *Hensen's node*) and in all cases it coincides with the place where embryonic cells pass through in the invagination process that gives origin to the dorsal cord. All vertebrates belong to the phylum *Chordata* precisely because the dorsal cord is a defining feature of that phylum, and this means that an organizer is present in all vertebrate embryos and is responsible for their body plan.

## 6.4 Determination and Cell Memory

Most of the differences that arise in embryonic development are generated by *inductions*, the processes where a group of cells sends off signals that induce other cells to make changes in the synthesis of their proteins and acquire in this way new histological characteristics. Induction, on the other hand, requires that a difference between inducer and induced tissue already exists, and is therefore a mechanism that can operate only from a certain point of development onwards. The earlier differences, it turned out, can be generated in two ways.

One is the so-called *mosaic determination*, a mechanism where the cytoplasm of the oocyte contains an asymmetric distribution of molecules that have the ability to determine the histological fate of the cells that acquire them. In *Drosophila*, for example, the distinction between sexual and somatic cells is due to the *pole plasm*, a substance that is deposited by the mother at the posterior end of the egg. The cells that receive molecules from the pole plasm become sexual cells and are potentially immortal, whereas all the others become somatic cells and are destined to die with the body.

The second mechanism is called *positional determination* because the cells are instructed to synthesize different proteins by the *position* that they happen to occupy.

In mammals, for example, the divisions of the fertilized egg produces a ball of cells where some occupy internal positions while others remain on the surface, and this is enough to create a lasting difference between them. All surface cells are committed to become part of the future placenta and only some internal cells maintain the potential to form an embryo.

The differentiation of the embryonic cells, in short, can be achieved by a variety of mechanisms, but all of them have one outstanding feature in common: the characteristics acquired by a cell are transmitted to *all* its descendants in the growing organism. In the course of development, in other words, the cells must not only *become* different but also *remain* different, and this is a formidable task.

When a cell divides by mitosis, its whole internal organization is dismantled in order to allow for the formation of the mitotic spindle, the transport of the chromosomes to opposite sites, and finally the split of the whole system into two daughter cells. But once these new cells are formed, how do they reconstruct the differentiated state of the mother cell? Do they have to be instructed by new inductions or do they somehow *remember* the original organization they came from?

The solution to this problem was again provided by Spemann with micro-transplants. In amphibians, for example, there is a very early stage where all cells on the surface are undifferentiated but occupy two distinct regions, one destined to become skin and the other nervous tissue. Spemann transplanted small pieces of tissue from the future brain-region to the future skin-region and vice versa and observed what happened. He found that the cells can change their histological fate (skin cells, for example, can become nerve cells) if they are transplanted *before* a critical period, but are totally unable to do so if the transplant takes place *after* that period. This means that there is a crucial period of development in which *something* happens that decides the histological fate of the cell and of all its descendants. That something was called *determination* and represents a *histological memory* for generations to come (Spemann 1938).

Other experiments proved that determination does not normally take place in a single step but in stages, and that the number of these stages varies from one tissue to another. What is most impressive, in all cases, is the extraordinary *stability* of determination. The process takes only a few hours to complete in the embryo and yet it leaves permanent effects in all descendant cells for the rest of the organism's life. The state of determination, furthermore, is conserved even when the cells are grown *in vitro* and divide many times outside the body. When brought back *in vivo*, they express again the properties of their determination state as if they had never 'forgotten' that experience (Alberts et al. 2007).

## 6.5 Self-Regulating Embryonic Fields

The fertilized egg divides into cells, called *blastomers*, that give origin first to a compact sphere, the *morula*, and then to a larger sphere with an inner cavity, the *blastocyst*. With microsurgery techniques it is possible to remove the morula, perform on it a prefixed set of operations, and then re-implant it in the natural mother or in a surrogate mother. This technology was developed to test the potentialities of the early embryonic cells, and in particular the claim advanced by Hans Driesch in the 1890s that isolated blastomers can develop into complete organisms (Driesch 1893, 1894). This amounted to saying that in early embryos a part can reconstruct the whole, an astonishing property that became known as *self-regulation.*

From the 1950s onwards, various experiments have proved that blastomers can indeed be removed from a morula and still maintain the ability to develop into normal embryos (Seidel 1952; Kelly 1977). Other experiments produced a complementary result: when two morulae are fused together and are re-implanted in a surrogate mother the resulting giant morula still develops into a normal creature (Tarkowski 1961; Mintz 1962).

Early embryos, in short, have two extraordinary characteristics: (1) their cells are *totipotent* because each of them has the potential to generate a complete organism, and (2) the embryos are *self-regulating systems*, because a part can reconstruct the whole and an excess of parts can be compensated.

The evidence, furthermore, shows that the blastomers loose their totipotency after a few cell divisions and simultaneously acquire new potentialities. This is expressed by saying that the embryo is subdivided into *embryonic fields* that increase in number but have more restricted potentialities. A clear example of this process is illustrated by what happens in human twins.

Monozygotic twins are formed when the blastomers that derive from a fertilized egg are accidentally split into two groups each of which has the potential to generate a complete embryo. The key point is that the extra-embryonic structures in which the embryo grows change their potentialities at prefixed stages of development and this has direct consequences for the twins (McLachlan 1994).

In the first three days, the extraembryonic cells have the potential to form a chorion (the embryonic part of the placenta) and an amnion, and if the split occurs in this period each twin has its own chorion and its own amnion (the twins are *dichorial and diamniotic*). Between the third and the eight day, the extraembryonic cells have lost the potential to form a chorion but maintain the potential to form an amnion, and if the split occurs in this period the twins grow in the same chorion but each in a separate amnion (the twins are *monochorial and diamniotic*). Between the eight and the fourteenth day, the extraembryonic cells have also lost the potential to form an amnion and the resulting twins grow together in the same chorion and in the same amnion (the twins are *monochorial and monoamniotic*).

Human twins are no longer formed after the second week of development because the potential to form all embryonic tissues is lost at the beginning of the third week, when the three germinal layers (ectoderm, mesoderm and endoderm) are formed and the embryonic cells acquire only one of these three potentialities. After that stage, the cells continue to differentiate and the germinal layers are further subdivided into new embryonic fields. These increasingly acquire the characteristics of *organ-forming-areas*, and the subdivision process continues until the forming-areas of all organs of the body are laid down. At this stage, no new embryonic field is formed and embryonic development is replaced by foetal development.

The overwhelming characteristic of the embryonic fields is *self-regulation*, the ability to compensate a defect or an excess of parts. In the case of the limb, for example, it has been shown that a normal limb is developed even when part of the limb-bud is removed or when an extra limb-bud is added to the original one (Harrison 1918; Saunders 1982; Gilbert 2006). A few animals conserve the ability to *regenerate* parts of their bodies even in adult life, but in most cases this ability is lost at the end of organogenesis, when the embryo becomes a foetus.

Embryonic development, in conclusion, is not only a cascade of inductions. It is a far more robust process because it has the ability to compensate countless accidental errors with the extraordinary mechanism of self-regulation, a mechanism whose understanding is still one of the great problems of biology.

## 6.6 Arranging Cells in Space

The various parts of a body may differ not only because they contain different tissues but also because their cells can be arranged in space in different ways. A right hand and a left hand, for example, or an arm and a leg, are made of the same tissues and yet their shapes are different, which means that in the course of development groups of identical cells can be arranged in different spatial configurations. But where do these spatial patterns come from? Are they the final result of a chain of embryonic processes, or must they be specified at the beginning by genetic instructions?

In 1894, William Bateson described a number of abnormalities due to mutations that he called *homeotic* because some parts of the body are developed normally but are situated in wrong places (Bateson 1894). *Antennapedia* mutations, for example, transform antennae into legs, thus giving rise to an insect with two legs sprouting from its head, whereas *bithorax* mutations transform the third thorax into a second one, giving the insect an extra pair of wings. This suggested that there are genes that control the *position* in space of the various parts of the body, but for a long time genetics was unable to prove their existence, because genes are recognized by mutations, and mutations of developmental genes normally bring development to a halt. This obstacle was overcome only thanks to some peculiar characteristics of the fruit fly *Drosophila*, and to the patient work of Edward Lewis at the California Institute of Technology (Lewis 1963).

Lewis noted that *Drosophila* larvae have 12 segments which can be individually recognized under the microscope because they carry hairs and denticles that have a specific morphology in every segment. Even if the insects were dying at the larval stage, therefore, the effects of their mutations could still be seen in the altered microscopic morphology of their segments. This made it possible to identify the loci of the *bithorax* mutations in the right arm of chromosome 3, and then the loci of the *Antennapedia* mutations in a separate cluster of genes situated on the same chromosome (Raff 1996).

Shortly afterwards came the demonstration that homeotic genes belong to a wider class of genes that control the entire body plan of the organism during embryonic development. Sander (1975), Garcia-Bellido (1975), Lewis (1978), Nüsslein-Volhard and Wieschaus (1980) and others were able to prove that the spatial organization of *Drosophila* embryos is built in stages under the control of three groups of genes: (1) axis genes, (2) segment genes, and (3) homeotic genes. Collectively, they have been called *Hox* genes because they have in common two important characteristics: one is that they are *positional* genes because they control not the components of a body but their *positions* in space; the other is that they contain a *homeobox*, a sequence of 180 nucleotides that encodes a protein domain of 60 amino acids that behaves as a DNA-binding domain (Gilbert 2006).

The first structure of the body plan is the head-to-tail axis, or *primary* axis, and in fruit flies the formation of this structure is initiated by maternal proteins deposited at the two extremities of the egg. The different concentrations of these proteins along the axis promote or inhibit the expression of the *Hox* genes in different ways, thus creating a first subdivision of the body into different zones. Genes of the *Hox* family, in other words, are expressed along the primary body axis in distinct zones, and the developmental characteristics of each zone reflect the combination of *Hox* genes that are activated or repressed in its cells.

Further interactions between *Hox* genes and proteins specify increasingly smaller zones and eventually divide the body into the series of segments that is the hallmark of insect organization. As soon as the fruit flies have been given their segments, the *Hox* genes start adding new features such as legs, wings, antennae and eyes. A gene called *Distal-less*, for example, initiates the development of the legs, whereas *eyeless* sets in motion the development of the eyes, and *tinman* starts the beginning of the heart (Carroll et al. 2001).

In embryonic development, in conclusion, there are two distinct types of genes at work: the genes that determine the histological characteristics of the cells (the *differentiation* genes), and those that determine their reference system, their positions in respect to the surrounding cells (the *pattern* genes).

## 6.7 The Unexpected from Molecular Embryology

All animal phyla that exist today were separated at the time of the Cambrian explosion, and have evolved along independent lines for more than 500 million years. In such an immense period of time, the ancestral bodies have been extensively modified to fit the needs of individual species. The adaptation to swimming, for example, has produced the same oblong body shapes in sharks, which are fish, and in whales, which are mammals. The forelimbs of birds and bats have been turned into wings by countless modifications of the basic limb bones that are common to all tetrapods. The evolutions of marsupials in Australia and of placental mammals in the rest of the world have taken place in complete separation, and yet the results show striking similarities: there are Australian marsupials that physically resemble true placental cats, mice, wolves, squirrels, moles, sloths and anteaters.

These results vividly illustrate the power of natural selection to produce countless changes in all organs of the body by modifying the genes that are responsible for their construction. This is why biologists concluded that from the Cambrian explosion onwards, natural selection has crafted and recrafted every single gene, including those that are responsible for the *homologous* structures that are derived from a common ancestor.

It was widely believed, in other words, that 500 million years of evolution have effectively erased any trace of *deep homology* in the genomes of distant animals. This conclusion was expressed by Ernst Mayr in no uncertain terms "... *the search for homologous genes is quite futile except in very close relatives*" (Mayr 1963, p. 609). This, in turn, implied that structures like eyes, limbs, and hearts, had to evolve independently in distant phyla. In the case of the eyes, for example, Salvini-Plawen and Mayr claimed that they evolved *de novo* some 40–65 times (Salvini-Plawen and Mayr 1977).

In the 1980s, however, it became clear that homologous genes do exist. The first step was the discovery that genes containing a homeobox exist not only in fruit flies but also in vertebrates, and most of them are true *Hox* genes because they have a *positional* function – they determine not the *components* of an organ but the *position* where the organ should be (Scott and Weiner 1984; McGinnis et al. 1984).

Then came the discovery that many *Hox* genes have equivalent functions in vastly distant animals such fruit flies, mice, and humans. Mutations in the gene *eyeless* in fruit flies, for example, cause eye abnormalities as do mutations in the gene *Pax-6* in vertebrates, thus suggesting a deep genetic homology between them. The conclusive proof came from the discovery that eyes can be artificially induced in fruit flies by *Hox* genes obtained from mice (Quiring et al. 1994; Halder et al. 1995; Gehring 1996).

The eyes of flies and mice are about as different as any two organs could be. They are definitely *not* homologues structures, and yet the *Hox* genes that position the eyes on the head are the same in the two animals, and apparently in all other animals. The eyes of flies and mice, in other words, are not homologous, but the genes that tell those animals where to place their eyes, are.

This was followed by discoveries that revealed the existence of homologous genes in many other organs. The *Hox* genes that initiate limb development in vertebrates, for example, are homologues to the *Distal-less* genes that operate in fruit flies, and the development of the vertebrate heart is initiated by a gene that is homologous to the fruit fly gene *tinman* (Carroll 1995; Carroll et al. 2001).

The study of the genes that operate in embryonic development, in short, has falsified one of the major predictions of population genetics – the alleged impossibility of *deep genetic homology* – and has given origin to the new research field of *Evolutionary Developmental Biology*, familiarly known as *EvoDevo*. The purpose of *EvoDevo* is uncovering the links that exist between embryonic development and evolution, and accounting for the very existence of homologous genes: how is it possible that many genes have been highly conserved in all animal phyla against the predictions of population genetics?

## 6.8 The Classification of Animals

Linnaeus (Carl von Linné) wrote *Systema Naturae* as a means of introducing order in the classification of all living creatures, thus revealing the 'Plan of Creation' that is hidden in Nature (Linnaeus 1758). In the case of animals he adopted a classification system based on five groups or *taxa* (species, genus, order, class and kingdom) but underlined that only 'species' and 'genus' are *natural* categories, whereas the other three are human abstractions. This is why he adopted a *binomial* terminology in which every living group is identified by two Latin names written in italics, the first indicating the genus and the second the species. The classification system was later extended by adding the taxa 'family' and 'phylum', but the binomial terminology has been conserved ever since. What has deeply changed in the field of classification, or *taxonomy*, is the *biological purpose* of the project, a purpose that after Linnaeus has undergone three substantial modifications.

The first was caused by the discovery of the body plans by Cuvier and von Baer. The existence of a common underlying structure in all vertebrates suggested the existence of an *archetype* that guides the building of all vertebrate bodies. The theory of the archetypes was first proposed by Goethe in plants (Goethe 1790) and later extended to animals by Geoffroy Saint-Hilaire (1818) and Richard Owen (1848). In that framework, the purpose of classification was the description of all archetypes that exist in Nature and of the *laws of forms* that derive from them.

The second great change in taxonomy came from Darwin (1859) in one swift stroke. Darwin pointed out that the existence of homologous organs in distant animals is much more naturally explained by the idea that they descend from a common ancestor. The attention shifted from archetypes to ancestors, and classification acquired an entirely new purpose: it became the project of reconstructing genealogic trees.

The third great innovation in taxonomy came about a century after Darwin, when Willi Hennig developed the principles and the techniques of *cladistics* (Hennig 1966). Henning insisted that classification must be based on genealogy, as Darwin said, and argued that true phylogenetic groups, or *clades*, are defined by shared features that are derived from a common ancestor and are exclusive to the group in question (features that are technically known as *synapomorphies*).

The protein *collagen*, for example, is found in all animal species and in no other creatures, a result which indicates that animals received that protein from a common ancestor. The presence of collagen, however, is hardly enough to *define* what animals are, and the same is true for many other biochemical, morphological and physiological features. In the end, all these features are the result of genetic instructions, so it is at the level of the genes that we should look for a proper answer.

This is the essence of the proposal that was put forward in 1993 by Jonathan Slack, Peter Holland and C.F. Graham. They underlined that in all animals there is a specific pattern of *Hox* genes that is expressed at the phylotypic stage of embryonic development, when the body plan of a phylum is most visible. That pattern, which they called *zootype*, defines the body plan in all animal phyla and can be regarded therefore as the defining feature (the *synapomorphy*) of kingdom Animalia (Slack et al. 1993).

## 6.9 The Cambrian Explosion

In the 1830s, Roderick Murchison discovered in the Cumbria region of Northwest England the greatest discontinuity ever found in the fossil record, the so-called *Cambrian explosion.* He found the geological stratum that contains the very first *visible* fossils of Earth's history. Whilst all lower strata were, to the naked eye, completely devoid of fossils, in the Cambrian one could see the fossilized remains of creatures that unmistakingly exhibited the sophisticated structures of highly developed metazoa (Murchison 1854).

For a long time, the beginning of the Cambrian has been associated with the appearance of *trilobites* (its most characteristic animals), but then it was found that these arthropods were preceded by the so-called *small shelly fossils*, minute animals with tiny shells that appeared – again 'suddenly' – in a lower Cambrian stratum that was called *Tommotian* (from the name of a Russian site). For a number of years the Tommotian marked the beginning of the Cambrian but even this conclusion had to be abandoned. The rocks immediately below it can also be included in the Cambrian but are practically devoid of *small shelly fossils*, and show instead a great number of *trace fossils*, that is to say tunnels and burrows that were undoubtedly excavated by small animals. Trace fossils do not allow us to reconstruct the animals that made them, but we can still divide those animals into groups, and even into distinct taxa. The most characteristic taxon was named *Phycoides pedum*, and eventually an international convention established that the appearance of *Phycoides* trace fossils marks the beginning of the Cambrian (Brasier et al. 1994).

Palaeontology, in conclusion, has discovered that in Cambrian times there have been not one but three different 'explosions' of animal life: the first documented by the *trace fossils*, the second which left behind *small shelly fossils*, and finally the classical explosion dominated by the *trilobites.* It must be added that Cambrian life was preceded by the so-called *Ediacara fauna*, a vast assembly of soft-bodied animals whose body plan do not correspond to known animal phyla and may well represent a failed evolutionary experiment (Glaessner 1983; Seilacher 1992).

The actual dating of the Cambrian and Precambrian rocks with radioactive methods has long been fraught with technical difficulties, but today the results are fairly trustworthy and great variations are no longer expected from future measurements (Des Marais 1997). This allows us to say that the Ediacara fauna flourished between 580 and 560 million years ago, whereas the Cambrian period started around 545 million years ago, and lasted some 40–45 million years (Conway Morris 1993).

What is most important is that the geological strata that contain *small shelly fossils* and those that contain *trilobites* (in particular the *Chengjiang* site and the

*Burgess Shale*) have a time range of 2–5 million years. The explosion of the *small shelly fossils* and the classical explosion dominated by *trilobites*, in other words, did not exceed 5 million years, a period that is extremely short from a geological point of view.

Another outstanding discovery has come from the comparative anatomy of the Cambrian fossils, whose studies have proved that the body plans of virtually all known animals were invented in Cambrian times (Gould 1989; Conway Morris 1998). The Cambrian explosion was traditionally defined as the appearance of the first skeleton-bearing metazoa, but now we can characterize it with a vastly more general event and say that it was *the geologically sudden appearance of all known animal phyla*. After the Cambrian explosion, in other words, many lower animal taxa came into being – new classes, new orders, new families, new genera and countless new species – but our planet has never again seen a new phylum.

## 6.10 The Origin of Animals

Most animals have an embryonic development based on three germs layers (ectoderm, endoderm and mesoderm), and for this reason are called *triploblasts*, but some develop from two germ layers (*diploblasts*) and a tiny minority from one layer only (*monoblasts*). The number of germ layers has an outstanding importance because it determines the overall symmetry of the body. A single germ layer produces organisms which have no body axis and no symmetry (the sponges); two germ layers build animals with one body axis and radial symmetry (the *radiata*, i.e., corals, polyps, jellyfish, etc.), whereas three germ layers give origin to animals that have three axes and a bilateral symmetry, i.e., bodies with a right side and a left side (the *bilateria*, i.e., vertebrates and invertebrates) (Tudge 2000).

The molecular data strongly suggest that animals are a *monophyletic* group, and we have therefore the problem of understanding how they evolved from a common ancestor. To this purpose, let us keep in mind that multicellular organisms can be generated by two distinct processes. One is the aggregation of free-living single cells that come together and form a *colony* (this is what takes place, for example, in slime moulds, giant kelps, seaweeds and myxobacteria). The other is the division of a cell into daughter cells that remain tightly bound together and form a *clone* of genetically identical cells. In the first case multicellularity is obtained by *colonial aggregation* and in the second case by *embryonic development*.

Colonial organisms can differentiate into distinct tissues, and this means that the first two key features of embryonic development – cell differentiation and cell memory – were probably invented by the ancestor of all multicellular organisms. The third key feature, instead, is *cell pattern*, the ability to form specific three-dimensional bodies, and this is present only in animals that have a body plan, i.e., in diploblasts and triploblasts. This means that the key event that set in motion the evolution of embryonic development was the invention of a body plan, more precisely the invention of the axes of symmetry that define a body plan.

It has been shown that *Hox* genes have a key role in the determination of the primary body axis in both diploblasts and triploblasts (Bayascas et al. 1998; Ryan et al. 2007) and this implies that they were present in the common ancestor of those animals. It has also been shown that the body plan of the common ancestor evolved in steps in its descendants. This is suggested, among other things, by the fact that the Ediacara fauna was entirely made of diploblastic organisms (Seilacher 1992) and triploblastic animals apparently came into being only in the Cambrian explosion (Conway Morris 1993).

The triploblastic body plan that defines the vast majority of past and present animals, in other words, appeared in Cambrian times after a long period of evolution, and the *Hox* genes had a key role in that event. We need therefore to take a closer at the relationship that exists between *Hox* genes and body plans.

## 6.11 The *Hox* Codes

In 1979, David Elder proposed a model that was capable of accounting for the regularities that exist in the bodies of many segmented worms (annelids). The segments of these animals are often subdivided into *annuli* whose number varies according to a simple rule: if a segment contains *n* annuli, the following segment contains either the same number *n* (*repetition*) or *n* plus or minus 1 (*digital modification*).

Elder noticed that this type of rules is known to the designers of electronic circuits as a *Gray code*, a code that is *binary* (because it employs circuits that have only one of two states), *combinatorial* (because its outcomes are obtained by combinations of circuits) and *progressive* (because consecutive outcomes must be coded by combinations that differ in the state of one circuit only). The results obtained with these rules describe with great accuracy what is observed in segmented worms, and Elder proposed therefore that the body plan of these animals is based on a *combinatorial code* that is a biological equivalent of the Gray code. He underlined in particular that the coding principle cannot be the classical "one gene-one pattern", but "one combination of genes-one pattern" and for this reason he called it *epigenetic code* (Elder 1979).

After the discovery of the *Hox* genes, it became increasingly clear that they are used in many different permutations, according to a combinatorial set of rules that became known as *Hox code*. The term *Hox* code was introduced independently by Paul Hunt and colleagues (1991) and by Kessel and Gruss (1991) to account for the finding that the individual characteristics of the vertebrae are determined by different combinations of *Hox* genes. Later on, it was found that this is true in most other organs and it became standard practice to refer to any combination of *Hox* genes as a *Hox* code. The epigenetic code proposed by Elder, in particular, is a *Hox* code because it is *Hox* genes that are responsible for the body plan of the segmented worms.

It must be underlined that the *Hox* genes can be used in different combinations not only in various parts of a body, but also in different stages of embryonic development. At the phylotypic stage, for example, the *Hox* genes specify characteristics of the phylum, whereas in later stages they determine characteristics at lower levels of organization. There is, in short, a hierarchy of *Hox* gene expressions, and therefore a hierarchy of *Hox* codes.

At this point, however, we have to face a key *definition* problem: is it legitimate to say that the *Hox* codes are true organic codes? More precisely, that they have the basic features that we find, for example, in the genetic code? An organic code is a *mapping* between two independent worlds and cannot exist without a set of adaptors that physically realize the mapping.

The *Hox* codes have been defined instead as *patterns of combinatorial gene expression* and do not require adaptors because a molecular pattern in one world is not a mapping between two independent worlds. We have therefore two different definitions of code, one based on mapping and the other on patterns, or sequences, and it is important to keep them separate because they have different biological implications.

## 6.12 The Codes of the Body Plans

The origin of animals gives us the same problem that we face in all major transitions: how did real novelties come into existence? In the case of the first animals the starting point was a population of cells that could assemble themselves in space in countless different ways, so how did they manage to generate those particular three-dimensional structures that we call animal bodies?

In principle, the number of arrangements that cells could form in space was unlimited, so it was imperative to make choices. These choices, or constraints, are the instructions that specify a body plan. More precisely the cells are instructed that their position is anterior or posterior, dorsal or ventral and proximal or distal *in respect to the surrounding cells*. These instructions are carried by genes, and consist in proteins that are referred to as the *molecular determinants* of the body axes (Gilbert 2006).

It must be underlined that the relationships of the body axes are between *cells*, and this means that they do not determine only the axes of the whole body, but also those of all its organs. In the hand, for example, the proximo-distal axis is the direction from wrist to fingers, the antero-posterior axis is from thumb to little finger, and the dorsal-ventral axis is from the outer surface to the palm of the hand. Right hand and left hand have different symmetries because their axes are one the mirror image of the other.

There is, in short, a multitude of axes in an animal body, and it turns out that many of them have the same molecular determinants. The products of the gene

*Sonic hedgehog* (*Shh*), for example, determine the dorso-ventral axis in the forebrain as well as the antero-posterior axis in the hand, which suggests that molecular determinants are mere labels.

The evidence has shown that this is true in all cases. There are countless determinants in animal species and all function as labels, or signs, in the sense that the links between them and body axes are not dictated by physical necessity. This means that they are based on conventional rules, on the rules of codes that can be referred to as *codes of the body plans*.

These codes create a mapping between two independent worlds – molecular determinants and body axes – and have therefore the quintessential feature that defines all organic codes. In this case, however, the mapping is not between two types of molecules, as in the genetic code. It is between molecules and body axes, and this inevitably makes the mapping extremely complex because body axes are relationships between *populations* of cells. This implies that the adaptors of the body plan may be difficult to identify, but the conclusion does not change. Mapping is the defining feature of the organic codes and the codes of the body plans create real mappings, so they are real organic codes. It will be noticed, furthermore, that they are more general than the *Hox* codes because in addition to *Hox* genes they include all the other components that are necessary to create a mapping bridge between molecular determinants and body axes.

From an evolutionary point of view, the codes of the body plans were instrumental to the origin of the animal phyla as the genetic code was instrumental to the origin of life, and represent another example of the close relationship that exists between organic codes and the great events of macroevolution. This idea has at least one outstanding consequence because it accounts for the fact that all animal phyla appeared in a geologically short period of time. The evolution of the individual rules of a code can take a long time, but the origin of a new code corresponds to the appearance of a *complete* set of rules and from a geological point of view this can be a quick event.

## 6.13 The Logic of the Embryos

To a large extent, it is legitimate to say that modern embryology has rediscovered the theories of epigenesis and immanent design proposed by Aristotle. Epigenesis is the classical idea that new structures appear in stages, whereas the modern version of immanent design is the idea that a genetic programme of development is present in the fertilized egg. This may give the impression that we now understand, at least in principle, the *logic* of embryonic development, and all that remains to be done is fitting in the details. In *The Problems of Biology*, however, John Maynard Smith has lucidly sounded a note of caution against this attitude:

> It is popular nowadays to say that morphogenesis (that is the development of form) is programmed by the genes. I think that this statement, although in a sense true, is unhelpful.

> Unless we understand how the program works, the statement gives us a false impression that we understand something when we do not ..... One reason why we find it so hard to understand the development of form may be that we do not make machines that develop: often we understand biological phenomena only when we have invented machines with similar properties .... and we do not make 'embryo' machines. (Maynard Smith 1986, pp. 99–100)

Maynard Smith's point can also be expressed in another way: embryonic development is a process that increases the complexity of a living system in a *convergent* way, but we do not know how to build machines that produce a convergent increase of complexity. It must be added, however, that we do have an algorithm that describes how such a goal can be achieved.

It is an algorithm developed for the reconstruction of two-dimensional images from one-dimensional projections, where it is known that a reconstruction is possible only if the projections contain as much information as the examined structure, a condition which implies that their number must not be inferior to a theoretical minimum (Barbieri 2003). A reconstruction from *incomplete* projections is equivalent to a convergent increase of complexity, and no algorithm can achieve it is a single step. It has been shown, however, that that goal can be achieved by a multi-step procedure, provided that (1) a number of 'memory' structures are built in parallel in order to store what takes place during the reconstruction, and (2) a number of codes are employed in order to transfer information from the memory space to the reconstruction space (Barbieri 2003).

This model is in no way a description of what happens in embryonic development, and yet it gives us two important general concepts because it suggests that (1) there can be no convergent increase of complexity *without memories*, and (2) there can be no convergent increase of complexity *without codes*.

This amounts to saying that memories and codes are essential ingredients in *any* convergent increase of complexity, and this makes us understand why they are present in embryonic development. We have seen, for example, that the histone code creates molecular patterns that amount to biological memories and that the codes of the body plans are largely responsible for the three-dimensional structures of the embryos.

What is emerging, in conclusion, is the idea that organic codes are key players in embryonic development but we still have a rudimentary knowledge of them and probably this is why we do not yet fully understand what goes on in a developing embryo. The important point, at any rate, is that the organic codes show the way forward, like an Ariadne's thread that allows us to navigate in the maze of complexity. Especially in the case of the most complex of all novelties created by embryonic development: the brain and the mind.

# Chapter 7
# Brain and Mind

An organism has mind when it has feelings and instincts, in a word when it has *first-person* experiences. The origin of mind was therefore the origin of *subjectivity*, the event that transformed some living objects into living *subjects*. There is a large consensus today that mental events are produced by brain events. More precisely, it is widely acknowledged that mind is made of higher-level brain processes, such as feelings and instincts, that are produced by lower-level brain processes such as neuron firings and synaptic connections. We need therefore to understand *how* does the brain produce the mind, and in particular *what is the difference* between them.

This chapter presents a new idea about the brain-mind relationship. The idea that there has been a nearly universal neural code at the origin of mind as there has been a nearly universal genetic code at the origin of life. This amounts to saying that there is a *natural divide* between mind and brain that is somewhat similar to the natural divide that exists between life and matter.

The key event in the origin of life was the appearance of *proteins*, and the key event in the origin of mind was the appearance of *feelings*. The first point that must be underlined is that there is an abyssal difference between proteins and feelings. Proteins are *space-objects*, in the sense that they act in virtue of their three-dimensional arrangements in space, whereas feelings are *time-objects* because they consist in processes, in flowing sequences of neural states. The same is true for their components. Proteins are assembled from smaller space-objects such as amino acids, whereas feeling are assembled from lower level brain processes such as neuron firings.

Despite this overwhelming difference, however, proteins and feelings have in common the fact that they are both the result of *manufacturing* processes. Proteins come into being when amino acids are joined together into sequences, and feelings come into being when neuron firings are assembled into first-person experiences. Proteins, in other words, are manufactured space-objects whereas feelings are manufactured time-objects. This is the *code theory of mind*, the idea that mind is produced by a manufacturing process based on a neural code as proteins are produced by a manufacturing process based on the genetic code (Barbieri 2006b).

© Springer International Publishing Switzerland 2015

M. Barbieri, *Code Biology*, DOI 10.1007/978-3-319-14535-8_7

This idea will be developed in stages in this chapter, and to that purpose we shall take a look at the main events that occurred in the evolution of the brain, starting from the appearance of the first neural cells.

## 7.1 Evolving the Neuron

All eukaryotic cells contain organelles that perform distinct functions and it is likely that this division of labour had a role in the evolution of different organs in the first animals. Some of their cells, for example, could preferentially express the genes of signal transduction and become the precursors of the future *sense organs*. Other cells could preferentially express the genes of locomotion, thus becoming the precursors of the future *motor organs*. A third type of cells could establish a link between them and prefigure in this way the future *nervous system* because this system is, by definition, a bridge between sense organs and motor organs. The point that must be underlined is that the cells of the nervous system have two key characteristics that evolved by modifying pre-existing structures.

The first major feature of the neuron is the ability to communicate with other cells by chemicals that are released from vesicles at points of close contact between their cell membranes (the *synapses*). It is those vesicles that provide the components of the brain signalling system, but they did not have to be invented from scratch. They evolved from the standard vesicles that exist in all eukaryotic cells and are routinely used for transporting molecules across the cell membrane.

The second great feature of the neuron is the ability to transmit electrical signals, and this too was obtained by a modification of pre-existing structures. Life was born in the sea, and even the organisms that invaded the land could do so only by carrying with them an *internal sea* that enabled their cells to continue living in water. This mother liquid that surrounds every cell has values of pH and osmotic pressure that are still similar to those of sea water, and likewise contains high concentrations of sodium and potassium ions. It turns out, however, that the concentrations of these salts inside the cell are totally different from those of the surrounding liquid. Sodium ions produce very high osmotic pressures and the cells are continually pumping them out with a battery of ion pumps and simultaneously importing potassium ions, whose osmotic pressure is much lower than sodium. The overall result is that there is a constant flux of ions across the cell membrane that takes place either by active transport (ion pumps) or by passive diffusion (ion channels).

The transport of ions across the cell membrane is also due to the fact that the interior of the cell is electrically negative in respect to the outside because most of the macromolecules that are trapped inside carry negative charges. The combination of this electrical asymmetry with the electric currents produced by ion pumps and ion channels leads to a stationary state characterized by an electrical difference across the cell membrane that is referred to as *membrane potential*.

This potential exists in all cells and any perturbation of it produces an electric pulse that is known as *action potential*. An electrical stimulus, for example, can open a sodium channel and let in a flux of positive ions that rapidly change the local value of the membrane potential. Such a change, however, is confined to a very small region under the cell membrane and can be propagated to other regions only if the membrane contains many other sodium channels at a close distance from each other. All cells, in short, have ions pumps and ion channels, but only an *uninterrupted chain of sodium channels* can propagate an action potential along the cell membrane. That was the key novelty that allowed a vesicle-producing cell to transmit electrical signals at long distances and become a *neuron*.

Chemical-releasing vesicles, ion pumps and ion channels, in conclusion, were all invented by free-living single cells during the first 3 billion years of evolution, and did not have to be rediscovered. All that was required for the origin of the neuron was a new way of organizing them in space and time.

## 7.2 The Intermediate Brain

The nervous system is made of three types of neurons: (1) the *sensory neurons* transmit electrical signals produced by the sense organs, (2) the *motor neurons* deliver electrical signals to the motor organs (muscles and glands), and (3) the *intermediate neurons* provide a bridge between them.

In some cases the sensory neurons are directly connected to the motor neurons thus forming a *reflex arch*, a system that provides an automatic stimulus-response reaction known as *reflex action*. Intermediate neurons, in other words, can be dispensed with, and a few animals do manage without them. It is a fact, however, that most animals have intermediate neurons, and what we observe in evolution is that brains increased their size primarily by increasing the number of these neurons. The evolution of the brain, in other words, has largely been the evolution of the *intermediate brain*.

It is known, today, that many neural processes are unconscious, and the intermediate brain is divided therefore into a *conscious* and an *unconscious* part. But when did this split occur? When did consciousness appear in the history of life?

Here, unfortunately, we come up against the difficulty that consciousness is too vast a category. It is associated with feelings, sensations, perceptions, emotions, instincts, thought, free will, ethics, aesthetics and so on. It is likely, however, that these entities appeared at different stages in the history of life and the origin of consciousness was probably associated with only a few of them, with something primitive and universal, something that even very simple animals have. William James, one of the founding fathers of modern psychology, pointed out that *feelings* are probably the most universal of all conscious processes, and it seems legitimate therefore to assume that consciousness came into being when the primitive brain managed to produce them (James 1890).

The first brains were probably a collection of reflex arches (as it is still the case, for example, in polyps and sea anemones) and it is likely that the first intermediate neurons evolved as a mere prolongation of those arches. Their proliferation was favoured simply because they provided a useful bridge between sensory neurons and motor neurons. Once in existence, however, they could start exploring other possibilities.

The first contribution of the intermediate neurons was probably the development of a multi-gated reflex-arch system. The behaviour of an animal must take into account a variety of clues from the environment, and to this purpose it is useful that a sense organ delivers signals to many motor organs and that a motor organ receives signals from many sense organs. This inevitably requires multi-gated connections between sensory inputs and motor outputs, and that probably explains why the intermediate brain had such a great evolutionary success.

In addition to *transmitting* electrical signals, however, the intermediate neurons could do something else. They could start *processing* those signals, and that opened up a whole new world of possibilities. In practice, the processing evolved into two great directions. One was the formation of neural networks that provided a sort of *automatic pilot* for the body. The other was the generation of *feelings*.

The first processing was totally unconscious and was carried out by a component of the intermediate brain that here is referred to as *cybernetic brain*. The second processing gave origin to another major component of the intermediate brain that here is referred to as *feeling brain*. The intermediate brain, in short, evolved from a primitive reflex-arch system and developed two distinct types of neural processing, one completely unconscious and the second controlled by consciousness. But why *two* types of processing? Why develop feelings if a cybernetic brain can work perfectly well without them?

## 7.3 The Instinctive Brain

A cybernetic brain can register changes in the environment and react with appropriate responses, so there does not seem to be any need for a feeling brain. We should keep in mind, however, that a cybernetic brain is a bridge between sense organs and motor organs that provides a *continuous* chains of reactions between inputs and outputs. This means that all operations of a cybernetic brain are linked together in a physically continuous sequence, and the initial input is invariably a signal from the outside world. An animal which has only a cybernetic brain, in other words, is virtually *a puppet in the hands of the environment*. A feeling brain, instead, is a system where the orders to act come from within the system, not from without. An animal with feelings takes decisions on the basis of its own motivations, and has therefore a certain degree of autonomy from the environment. But does such autonomy have an evolutionary advantage?

In circumstances when there is no food or no sexual partner in the immediate surroundings, a cybernetic animal would eventually stop eating and mating, whereas a feeling animal would embark in long journeys of exploration, well beyond its visible surroundings and even in the absence of external stimuli. An internal drive to act irrespective of the circumstances, in short, can have a strong survival role, and that is probably why most animals evolved both a cybernetic brain *and* a feeling brain.

It must be underlined, however, that the evolution of a feeling brain required a major change in brain circuitry. A feeling brain acts on the basis of internal drives, and this means that it has the ability to generate its own electrical signals and send orders to the motor organs even without signals from the sense organs. This implies that the continuity between sense organs and motor organs provided by the cybernetic brain had to be *interrupted*, and the gap filled by a new bridge made of feelings. The brain, in short, did not simply *add* feelings to a pre-existing system. It physically broke the continuity of the cybernetic bridge and introduced a new bridge in between.

As a result, the intermediate brain, acquired two distinct control systems, one totally unconscious (the cybernetic brain) and the other based on consciousness (the feeling brain). A common factor between them is the fact that their reactions are often referred to as *instincts*. More precisely, cybernetic reactions can be described as *unconscious instincts* and feelings as *conscious instincts*. This is why the unity of cybernetic brain and feeling brain is referred to as *instinctive brain*.

It has long been assumed that the cybernetic brain is a servo-mechanism of the conscious brain, but this traditional idea had to be abandoned when Benjamin Libet proved that the unconscious brain activity that prepares the execution of a movement (for example flexing a finger) starts *before* one consciously takes the decision to make that movement (Libet 1985). The interpretation of Libet's classic experiments is still a matter of discussion, but the neurological data make it clear that the cybernetic brain and the conscious brain are two largely autonomous systems that somehow manage to work together as an integrated whole.

Before discussing these two systems, however, we need to take a look at the way in which the brain is processing the data from the world.

## 7.4 Making Mental Images

The human retina consists in three layers of neural cells. The first contains about 100 million *photoreceptor cells* (rods and cones) that react to light by producing electrical signals. These are sent to the *bipolar cells* of the second layer which in turn deliver signals to the one million *ganglion cells* of the third layer whose axons form the optic nerve. The 100 million signals of the photoreceptor cells undergo therefore a first processing on the retina that transforms them into one million pulses delivered via the optic nerve to the brain. Here the signals are sent to the midbrain and after the *optic chiasm* (where 50 % of the nerve fibers swap direction) are transmitted to the *visual cortex*, at the back of the head, where they are further processed by groups of *cortical cells* arranged in distinct *areas*.

It turns out that the operations performed in areas 17, 18 and 19 of the visual cortex maintain a certain topological coherence with the retina, in the sense that adjacent points in the retina are processed by adjacent points in the visual cortex. In area 17, furthermore, Hubel and Wiesel discovered that some cells react only to horizontal movements on the retina, other cells react only to vertical movements and still others to sharp edges (Hubel and Wiesel 1962, 1979). After areas 18 and 19, the visual inputs go on to other cortical areas, but the topological coherence with the retina is rapidly lost, probably because the information about spatial relationships has already been extracted.

These classic discoveries have shown that neural images are built *in stages*, and such a hierarchy can only have been the result of a long history, so let us take a brief look at the evolution of vision.

Some of the most primitive eyes are those of flatworms and are little more than clusters of photoreceptor cells that distinguish daylight from darkness. They can also detect the direction of a light source, a feat that allows flatworms to swim towards the dark and hide. But flatworm eyes do not have a lens and thus cannot form visual images of the surrounding objects.

The first camera-eye, with a lens that projects an image on the retina, probably appeared in fish. The fish retina already has a three-layered structure (rods and cones, bipolar cells and ganglion cells) and an optic nerve that transmits the visual inputs to the midbrain. In fish, however, all nerve fibers change direction at the optic chiasm, and the midbrain is the final destination of the visual inputs.

This primitive structure was substantially conserved in amphibians and reptiles, and it was only birds and mammals that evolved a more advanced design. In their visual system, only some fibers of the optic nerves cross direction at the optic chiasm and the final destination of the visual inputs moved from the midbrain first to the visual cortex and then to other regions of the neocortex. These changes went hand in hand with a gradual transition from an olfactory and tactile mode of life to a lifestyle where vision was acquiring an increasingly important role.

It has been found, furthermore, that the processing of visual information is fundamentally similar to the processing of olfactory, gustative, auditory and tactile information. All sense organs transform external messages into electrical signals that are sent to the cortex, where each sensation has its *primary* sensory areas. After that, the signals are delivered to the *secondary* areas and then to the *association* areas, where the different sense stimuli from any given object are integrated into a multisensory neural representation. Finally the neural signals are sent to the motor areas and become instructions to muscles and glands to take action.

## 7.5 Sensations and Perceptions

The brain does not merely *receive* information from the retina. It can literally *manipulate* it. When an object is approaching, for example, its image on the retina becomes larger, but the brain still perceives an object of constant size. When the head is moving, the image on the retina of a fixed external object is also moving, and yet the brain corrects that impression and concludes that the object is standing still. When the light intensity is lowered, the retinal image of a green apple, for example, becomes darker, but the brain compensates for that and decides that the apple has not changed its colour (Delbrück 1986).

These and many other results prove that what we 'perceive' is not necessarily what the sense organs have registered. *Perceptions*, in other words, are distinct from *sensations*. A sensation is what comes from the senses and has a specific physiological effect (colour, sound, smell and so on). A perception is what the brain decides to do of the information that comes from the senses.

All sense organs transform external signals into electrical pulses, and in this sense are comparable to the signal transduction receptors of the cell membrane that transform first messengers into second messengers. In both cases there is no necessary connections between external and internal signals and their transduction takes place according to coding rules. As there are signal transduction codes in the cell, in other words, there are *neural transduction codes* in the sense organs. It has been reported, for example, the existence of a *neural code for taste* in the gustatory system (Di Lorenzo 2000; Hallock and Di Lorenzo 2006), a combinatorial *odorant receptor code* in the olfactory system (Dudai 1999; Ray et al. 2006) and a *space code* in the hippocampus that provides an internal spatial map of the environment (O'Keefe and Burgess 1996, 2005; Hafting et al. 2005; Brandon and Hasselmo 2009; Papoutsi et al. 2009). In the intermediate brain, in short, there are neurons that are processing electrical signals with mechanisms that are, to all effects, the neural equivalent of the signal transduction processes that exist in every cell.

It must be underlined that the signals produced by the sense organs do not carry *organ-specific* instructions. They all consist in the same type of electrical pulses whatever is the organ they come from, and their sole function is to regulate the intensity and the frequency with which a prefixed sensation is expressed. This is because the neurons of the intermediate brain are committed to a specific reaction mode since the early stages of embryonic development, when their fate is fixed for life. A neuron that has been committed to the sensation of smell, for example, reacts with an olfactory activity to whatever electrical signal it receives, and in the body this is the right reaction only because it is receiving signals from the olfactory organs, via connections that were laid down in embryonic development.

All of which implies that the neurons of the intermediate brain are subject to two distinct types of cell determination. One is the process that differentiates the nervous system from all other tissues, whereas the other is the commitment to specialize in an organ-specific neural activity.

## 7.6 A Universal Neural Code

The term *neural code* is used fairly often in the scientific literature and stands for the unknown mechanisms by which the intermediate brain transforms the signals from the sense organs into subjective experiences such as feelings, instincts and sensations. The term, however, is potentially ambiguous, because it may indicate either a universal code or a set of rules that animals use to create their species-specific representations of the world. A similar ambiguity exists, for example, in the term 'language', which can indicate either a universal human faculty or the specific language that is spoken in a particular place. On this issue, therefore, let us first examine the indications that come from laboratory experiments.

Let us consider, for example the transformation of mechanical stimuli into tactile sensations. Rats have mechano-receptors on the tip of their whiskers while we have them on the tip of our fingers, and there is no doubt that our tactile exploration of the world is different from theirs, but does that mean that we use a different neural code? Virtually all experiments performed so far suggest that the neural processes that transform mechanical stimuli into tactile sensations are the same in all animals, and this does suggest that there is a universal mechanism at work (Nicolelis and Ribeiro 2006; Nicolelis 2011). The evidence in question comes from animals with three germ layers (the triploblasts), but they represent the vast majority of animals, so let us concentrate on them. On what grounds can we generalize the experimental data and conclude that virtually all triploblastic animals have a universal neural code?

We know that all neural processing operations start from the electrical signals produced by sense organs, but we also know that the sense organs arise from the basic histological tissues of the body, and these tissues (epithelial, connective, muscular and nervous tissues) are the same in all triploblastic animals. All signals that are delivered to the brain, in other words, are produced by organs that arise from a limited number of universal tissues, and represent therefore a limited number of universal inputs. But do we also have a limited number of universal outputs?

The neural correlates of the sense organs (feelings and instincts) can be recognized by the *actions* that they produce, and there is ample evidence that all triploblastic animals have the same basic *instincts*. They all have the imperative to *survive* and to *reproduce*. They all seem to experience hunger and thirst, fear and aggression, and all are potentially capable of reacting to stimuli such as light, sound and temperature. The neural correlates of the basic histological tissues, in short, are associated with the basic animal instincts and these appear to be virtually the same in all triploblastic animals.

What we observe, in conclusion, is a universal set of basic histological tissues on one side, a universal set of basic animal instincts on the other side, and a set of neural transformation processes in between. The most parsimonious explanation is that the neural processes in between are also a universal set of operations. And since there is no necessary physical link between sense organs and feelings, we conclude that the bridge between them can only be the result of a virtually universal *neural code*.

Sense organs and feelings, on the other hand, do not come full blown into existence. They are built in stages during embryonic development, so it is to their ontogenesis that we must turn our attention.

## 7.7 Mechanisms of Brain Development

The embryonic development of the vertebrate brain takes place in four stages. The first begins when a strip of ectoderm is induced to become neural tissue by the underlying mesoderm, and comes to an end when the newly formed neuroblasts complete their last cell division, an event which marks the *birth* of the neurons. This is a truly epochal event because everything that a neuron will ever do in its life is largely determined by the time and the place of its birth. Somehow, these two parameters leave an indelible mark in the newly formed neuron and become a permanent memory for it.

The second phase of neural development is the period in which neurons migrate from their birthplace to their final destination, a target that they 'know' because it is somehow 'written' in the memory of their birth.

The third phase begins when neurons reach their definitive residence. From this time onward, the body of a neuron does not move any more but sends out 'tentacles' that begin a long journey of exploration in the surrounding body. Any tentacle (a *neurite*) ends with a roughly triangular lamina (called *growth cone*) which moves like the hand of a blind man, touching and feeling any object on its path before deciding what to do next. The axons of motor neurons are the longest of such tentacles, and their task is to leave the neural tube and move into the rest of the body in search of organs that require nerve connections. This is achieved with an exploration strategy that takes place in two stages.

In the first part of the journey, the growth cones move along tracks provided by specialized molecules, with a preference for those of other axons (which explains why growth cones migrate together and form the thick bundles that we call *nerves*). The neurons do not have a geographical knowledge of their targets but this is compensated by an overproduction of cells, which ensures that some of them will actually reach the targets. At this point the second part of the strategy comes into play. The organs that need to be innervated send off particular molecules, known as *nerve growth factors*, that literally save the neurons from certain death. More precisely, neurons are programmed to commit suicide – i.e., to activate the genes of cell death, or *apoptosis* – at the end of a predetermined period, and nerve growth factors are the only molecules that can switch off this self-destruction mechanism. The result is that the neurons that reach the right places survive, and all the others disappear (Levi-Montalcini 1975, 1987; Changeaux 1983).

The fourth phase of brain development begins when the growth cones reach the target areas. At this point, some unknown signal instructs the axon to stop moving and to begin a new transformation. The growth cone looses its flat shape and generates a variety of thin long fingers that are sent off in various directions towards the surrounding cells. When a contact is established, the tips of the finger-like extensions expand themselves and become the round buttons of the *synapses*, the structures that specialize in the transmission of neurochemicals. This turns the neuron into a secretory cell and from that moment on the neuron is committed to a life of uninterrupted chemical communication with other cells.

The making and breaking of synaptic connections is the actual wiring of the nervous system and takes place with a mechanism that is based first on molecular recognitions and then on functional reinforcements. Each neuron generates an excess number of synapses, so the system is initially over-connected. The synaptic connections, on the other hand, are continuously broken and reformed, and only those that are repeatedly reconnected become stable structures. The others are progressively eliminated and in the end only the active synapses remain. This mechanism continues to operate long after birth thus providing the means to form new neural connections throughout the life of an individual.

## 7.8 Codes in Brain Development

Cell adhesion, cell death and cell differentiation are major tools of brain development, and in all of them we can document the presence of organic codes. Let us briefly examine a few examples.

1. *The adhesive code*
   In the 1940s, Roger Sperry severed the optic nerve of a fish and showed that its fibres grow back precisely to their former targets in the brain. More than that. When the eye was rotated 180° in its socket, the fish was snapping down rather than up at a bait placed above it, thus proving that the connections are extremely specific. This led Sperry (1943, 1963) to formulate the *chemoaffinity hypothesis*, the idea that neurons recognize their synaptic partners by millions of 'recognizing molecules' displayed on their cell membranes. The wiring of the brain is essentially accomplished by molecules that bridge the synaptic cleft and decide which neurons are connected and which are not. They function both as synaptic recognizers and synaptic glue, and recently it has been shown that cadherins and protocadherins are good candidates for these roles. Protocadherins, in particular, have an enormous potential for diversification because their genes contain variable and constant regions like the genes of the immunoglobulins. They could, therefore, provide the building blocks of a neural system that is capable of learning and memorizing, and, like the immune system, can cope with virtually everything, even the unexpected (Hilschmann et al. 2001). This suggests that the chemoaffinity hypothesis of Roger Sperry should be re-formulated in terms of coding rules. Rather than listing millions of individual molecular interactions, an organic code can generate an enormous diversity with a limited number of rules, and this is why various authors have proposed that the wiring of the nervous system is based on an *adhesive code* (Redies and Takeichi 1996; Shapiro and Colman 1999).
2. *The apoptosis code*
   Active cell suicide (apoptosis) is a universal mechanism of embryonic development, one that is used to shape virtually *all* organs of the body. The key point is that suicide genes are present in all cells and the signalling molecules that switch them on and off are of countless different types. This means that there are no necessary connections between the recognition of a signalling molecule and the activation of the suicide genes. They are two independent processes, and the only realistic solution is that the link between them is established by the conventional rules of an *apoptosis code*, i.e., a code that determines which signalling molecules switch on the apoptosis genes in which tissue. This

conclusion has been supported by various studies of apoptosis, in particular by those carried out by Basañez and Hardwick (2008) and by Füllgrabe et al. (2010).

3. *The transcriptional codes*
   Neurons arise from progenitor cells that line the lumen of the neural tube and acquire phenotypes that are specific for their site of origin. In the spinal cord, a gradient of morphogenetic molecules defines distinct domains of progenitor cells that express specific combinations of *homeodomain transcription factors*, and it is the combinatorial action of these factors that determines the expression of the genes of cell fate. It was shown that neural fate can be experimentally changed by altering the combinations of transcription factors, which strongly suggests that the determination of cell fate in the spinal cord is based on the rules of a *transcriptional code* (Jessell 2000; Marquard and Pfaff 2001; Ruiz i Altaba et al. 2003; Flames et al. 2007). In other parts of the nervous system, the specification of cell fate is not achieved by spatial patterning into progenitor domains, but rather by the temporal production of different cell types in a defined order. In this case too, however, there is evidence of a combinatorial code of transcription factors because it has been shown that there is a window of plasticity during which the fate of the neural cells can be experimentally altered (Shen et al. 2006; Osborne et al. 2008).

In brain development, in conclusion, we see at work a variety of processes that are based on coding rules, and this reinforces the idea that organic codes exist at all levels and are an integral part of life.

## 7.9 Theories on the Brain-Mind Relationship

Today the scientific theories that have been proposed on the relationship between brain and mind can be divided into three major groups.

1. The *computational theory* is the idea that lower-level brain processes are transformed into feelings and instincts by neurological processes that are equivalent to computations. Brain and mind are compared to the hardware and software of a computer, and mental activity is regarded as a sort of data processing which is implemented by the brain but is in principle distinct from it, rather like software is distinct from hardware (Fodor 1975, 1983; Johnson-Laird 1983). Some evidence does appear to support this conclusion. Roger Shepard, for example, has shown that mental images can literally be 'rotated' in the head. He asked people to tell whether or not two objects were identical, and found that the time required to give the answer was proportional to the angle by which the neural images of the two objects had to be mentally rotated (Shepard and Metzler 1971). This and various other results support the conclusion that the brain performs a sort of computing operations on neural entities.
2. The *connectionist theory* maintains that the brain is solving problems not so much by performing abstract computations but by actually 'fabricating' the

solutions with neural networks that operate with explorative strategies. The reference model, here, is the computer-generated neural networks that simulate the growth of synaptic webs in developing brains, and it has been shown that artificial neural networks are indeed capable of solving a wide variety of problems (Hopfield 1982; Rumelhart and McClelland 1986; Edelman 1989; Holland 1992; Churchland and Sejnowski 1993; Crick 1994). Neural computations and neural networks are eminently suited to describe the automatic operations of the unconscious brain, and let us not forget that these account for the greatest part of brain activity. The fact remains, however, that they do not require consciousness, and this is why a third school of thought has proposed that consciousness, or mind, is a faculty that 'emerges' from neural processes.

3. The *emergence theory* states that higher-level brain properties emerge from lower-level neurological phenomena, and mind is distinct from brain because any emergence is accompanied by the manifestation of new properties (Morgan Lloyd 1923; Searle 1980, 2002). This amounts to saying that mind is produced by the brain but is somehow distinct from it, because the properties of a system are not entirely reducible to those of its components (the properties of water, for example, are not accounted for by the properties of hydrogen and oxygen). The limit of this proposal is that mind becomes an 'epiphenomenon' of brain activity. It does represent something different but it does not require a mechanism. Once we have brain activity we automatically have emergent properties, so we do not need an explanation for them. Mind emerges from the brain as water emerges from hydrogen and oxygen.

A new version of the emergence theory is the idea that mind, or consciousness, is the result of *quantum effects*, an idea which has been given a variety of formulations (Marshall 1989; Zohar 1990; Stapp 1993). In this framework, consciousness becomes a new state of matter that requires very special circumstances to manifests itself, but once these circumstances exist the result is as automatic as a quantum phase transition. Again, consciousness would spring into existence spontaneously, as in emergence.

The mechanisms that operate in the brain, in conclusion, are neural signalling, neural computations and neural networks and none of them directly produces conscious experiences. Which means either that consciousness is an illusion or that another mechanism is at work in the nervous system.

## 7.10 The 'First-Person' Experiences

Feelings, sensations, emotions and instincts are often referred to as *first-person* experiences because they are experienced directly, without intermediaries. They make us feel that we *know* our body, that we are in charge of its movements, that we are conscious beings and that we live a personal life. Above all, they are quintessentially *private* internal states, and this makes it impossible to share them with other people.

The goal of science is to produce testable models of what exists in nature, and first-person experiences are undoubtedly part of nature, so we should be able to deal with them. At the same time, models are not reality ("*the map is not the territory*"), and what really matters is that they can be tested and improved. In our case, the problem is to build models that make us understand, at least in principle, how first-person experiences can be produced.

Let's take, for example, the case in which a toe is injured. We know that electrical pulses are immediately sent to the intermediate brain that processes them and delivers orders to the motor organs which spring the body into action. Here we have two distinct players: an observer system (the intermediate brain) and an observed part (the injured toe). It is the observer that gets the information and transforms it into the feeling of pain, but then something extraordinary happens. We do not feel the pain in the intermediate brain, where the feeling is created, but in the toe, where the injury took place. Observer and observed have become one, and it is precisely this collapse into a single feeling entity that generates a 'first-person' experience.

Something similar takes place when we receive signals from the environment, for example when we look at a tree in the background. In this case, an image is formed on the retina and electrical signals are sent to the intermediate brain. Again, there is a separation between the sender of signals (the retina) and the receiver (the brain) and yet we do not see an image on the retina, where the visual information is actually generated, nor in the brain, where it is processed and integrated. What we see is a tree in the outside world. This is again a first-person experience and again it is generated by a physiological process that somehow creates a short-circuit between observer and observed.

Signal transduction processes are continuously going on in the sense organs of the body and in the intermediate brain, and yet we are totally unaware of them. Consciousness is not the *direct* result of signal transduction. It is the result of additional operations that somehow bring together events that occur in the body and events that occur in the brain. This is an active *manipulation* of neural states, and we know that it is a manipulation because we also experience it in dreams. Consciousness, in other words, is *manufactured*, and any manufacturing process is necessarily based on coding rules.

What we call 'first-person' experiences, in brief, is nothing elementary, undifferentiated and indivisible. The exact opposite is true. They are the result of complex manipulations where highly differentiated cells act in concert to create a physiological short-circuit between body and brain, between senders and receivers of neural signals. This, incidentally, is why first-person experiences do not exist in single cells. They could evolve only in multicellular systems and their origin was a true macroevolution, an absolute novelty in the history of life. An absolute novelty that required, like many other macroevolutions, the origin of a new code.

## 7.11 The Code Theory of Mind

Proteins come into being when amino acids are joined together in a specific sequence, and feelings come into being when events that occur *separately* in the body and in the brain are somehow brought together into a single experience. Proteins, in other words, are manufactured molecules whereas feelings are manufactured neural processes. This is the *code theory of mind*, the idea that mind is produced by a manufacturing process based on a neural code as proteins are produced by a manufacturing process based on the genetic code (Barbieri 2006b).

We have seen that the evidence accumulated so far does support the existence of a nearly universal neural code (Nicolelis and Ribeiro 2006; Nicolelis 2011), but do we have some evidence also for the idea that mind is manufactured? Probably the best argument in favour of this idea is the existence of *dreams*: whatever they are, there seems little doubt that they are the result of a manufacturing activity.

The idea that mind is manufactured implies that the difference between mind and body is rather like the difference that exists between life and matter. Life evolved from matter and yet it is fundamentally different from it because entities like information and codes simply do not exist in the inanimate world.

Any biological system that manufactures objects according to the rules of a code is generating biological artifacts, and a world of artifacts is fundamentally different from a world of spontaneous objects precisely because it contains sequences and coding rules. It is the origin of artifacts, in short, that accounts for the appearance of genuine novelties in evolution.

The intermediate brain evolved as a means of processing the signals from the sense organs and most of its operations are carried out by neural computations and neural networks. The origin of consciousness, however, required the invention of a completely different mechanism and that was a true macroevolution because it brought into existence entities, such a feelings and instincts, that had never existed before.

According to the *code theory of the mind*, feelings and instincts are not produced by emergence, are not side-effects of neural networks, and are not the result of computations, but of real *manufacturing* processes. In this framework, in other words, feelings and instincts are manufactured *artifacts*, whereas according to the other theories they are *spontaneous products* of brain processes. This is relevant to the mind-body problem because the mind could not have *rules of its own* if it were made of spontaneous products, whereas the coding rules of artifacts can generate absolute novelties. The autonomy of the mind, in other words, is something that spontaneous brain products cannot achieve whereas brain artifacts can.

## 7.12 The Interpretive Brain

The feeling brain is the driving force of an animal, the main responsible for its ability to behave as an autonomous system. The cybernetic brain is essentially an automatic pilot, and it is this function that explains its enormous increase in the history of life. The greatest changes in animal evolution have taken place precisely in the cybernetic tools that animals evolved in order to provide the brain with increasingly sophisticated feedback mechanisms.

The *neural networks* are probably the most powerful of such tools. Their ability to create feedback loops allow them to produce a goal-directed behaviour, but they also have other outstanding properties. Neural networks have the ability to form *memories*, and a set of memories is the basis of *learning* because it allows a system to decide how to behave in any given situation by comparing the memories of what happened in previous situations (Hopfield 1982; Rumelhart and McClelland 1986; Holland 1992). A set of memories, in other words, amounts to a model of the world that is continuously updated and allows a system to *interpret* what goes on around it.

In real life, a system is necessarily formed by a limited number of memories whereas the world offers a potentially unlimited number of possibilities. Clearly, such a system can never be perfect, but it has been shown that neural networks can in part overcome this limitation by interpolating between discreet memories (Kohonen 1984). In a way, they are able to 'jump-to-conclusions', so to speak, from a limited number of experiences, and in most cases these 'guesses' are good enough for practical purposes.

This *extrapolation from limited data* is an operation that is not reducible to the classical Aristotelian categories of induction and deduction, and for this reason Charles Peirce (1906) called it *abduction*. It is a new logical category, and the ability to interpret the world appears to be based precisely on that logic.

An act of interpretation is a process that gives meaning to something, and this is, by definition, an act of semiosis. Interpretation, however, is a new form of semiosis because it is based not only on coding and decoding but also on abduction, and represents therefore a real novelty in the history of life. What is interpreted, furthermore, is not the world but *representations* of the world, and this means that interpretation can exist only in multicellular systems.

The origin of animals was a true macroevolution because it brought into existence a cognitive system based on feelings and instincts. Later on, another major transition allowed some animals to evolve a second cognitive system that gave them the ability to *interpret* the world. That system is a distinct part of the intermediate brain that can be referred to as the *interpretive brain* of an animal.

## 7.13 Three Brain Macroevolutions

The products of the brain are usually called feelings, sensations, perceptions, emotions, thoughts and so on, but it is useful to also have a more general term that applies to all of them, and to that purpose they are often referred to as *models* or *representations*. This amounts to saying that the intermediate brain uses the signals from the sense organs to generate *models* of reality. A visual image, for example, is a model built with the signals delivered by the retina, and a feeling of hunger is a model obtained by processing the messages sent by the sense detectors of the digestive apparatus.

The brain can be described in this way as a *modelling system*, a concept that has been popularized by Thomas Sebeok and has acquired an increasing importance in semiotics (Sebeok and Danesi 2000). The term was actually coined by Juri Lotman, who described language as the 'primary modelling system' of our species (Lotman 1991), but Sebeok underlined that language evolved from animal ancestors and should be regarded as a secondary or a tertiary modelling system.

Many authors prefer to describe the brain as a *cognitive system* rather than a modelling one, but even in this case we need to distinguish between primary, secondary and tertiary implementations, so it is important to be clear about them. Here we use those terms to indicate the cognitive (or modelling) systems that appeared at three different stages of evolution and gave origin to three distinct types of brain processing.

1. *The first cognitive system*
   This is the system that gave origin to sensations and consists in two great subsystems, because the sense organs deliver information either about the outside world or about the interior of the body. The first cognitive system makes therefore two types of models, one that represents the environment and one that carries information about the body. Jakob von Uexküll (1909) called these two worlds *Umwelt* and *Innenwelt*, names that express very well the idea that every animal lives in two distinct subjective universes. We can say therefore that *Innenwelt* is the model of the internal body of an animal and *Umwelt* is the model of its external world. The basic animal brain – the instinctive brain – came into being when the primordial intermediate brain split into feeling brain and cybernetic brain, and these started producing the conscious and unconscious instincts that apparently exist in all triploblastic animals (vertebrates and invertebrates).
2. *The second cognitive system*
   This is the system that allows a large number of animals to *interpret* the signals from the environment by using processes of abduction implemented by neural networks. The faculty of interpretation did not appear full blown but evolved in stages, and we can still see the descendents of creatures that represent intermediate levels in this step-by-step evolution. Snakes, for example, stop chasing a prey when it disappears from sight, whereas other animals deduce that the prey has temporarily been hidden by an obstacle and continue chasing it. Some can even follow its footsteps, which reveals a still higher level of interpreting power.

3. *The third cognitive system*
   This is language, a system that evolved only in our species and allowed it to build, as we will see in the next chapter, an entirely new world of symbolic objects that we call *culture*.

There have been, in conclusion, three major transitions in the evolution of the brain, and each of them gave origin to a new type of neural processing that was, to all effects, a new cognitive system.

# Chapter 8
# Origins of Language

The idea that man is different from animals is present in all cultures and is generally expressed by saying that only man has 'higher' faculties like consciousness, free will, morality and the creative power to produce art, religion, science, and poetry (together with torture, mass murder, and environmental disasters). Today we have a shorter explanation for all that. All we need to say is that only man has *language*. The rest is just a consequence of that one faculty, so it is the origin of language that we need to understand if we want to find out what made us human.

Anything in science, however, must be accounted for, including the sweeping generalization that we have just encountered. Are we really sure that only man has language? That animals do not have simpler forms of language? Granted that animals do not talk, they are certainly capable of communicating, often in highly sophisticated ways, so it seems reasonable to conclude that language is but an evolved form of animal communication. More powerful, yes, but not *qualitatively* different. This is possible, of course, but we should not take it for granted because genuine novelties did, occasionally, appear in the history of life.

This chapter argues that the origin of language was a true macroevolution, not a more elaborate form of animal communication. It will be shown, in particular, that the preconditions for language were created by a deep change in the foetal development of our species. More precisely, by a fetalization process that produced an extra-uterine phase of foetal development, and gradually extended it to the point that it became longer than the intrauterine one. This uniquely human form of development created a condition that is referred to as *cerebra bifida*, in some ways analogous to the condition of *cardia bifida* that is well known from laboratory experiments.

The main thesis of this chapter is that the extrauterine phase of human foetal development favoured the building of an entirely new cognitive system that evolved side by side the faculties that we have inherited from our animal ancestors. This model accounts for the fact that we have many faculties in common with animals and yet language is a cognitive system that did not evolve in other species.

© Springer International Publishing Switzerland 2015

M. Barbieri, *Code Biology*, DOI 10.1007/978-3-319-14535-8_8

## 8.1 Chomsky's Definitions of Language

In modern linguistics, any verbal system (English, Russian, Chinese, etc), is referred to as *external* language, whereas the faculty that is responsible for it is called *internal* language. In everyday life, the term 'language' is normally used in the first sense, whereas in academia and in scientific research it is mostly used in the second sense, so we need to keep in mind that there is a substantial distinction between external and internal language.

Another important distinction is between language and *speech*. Speech is the actual verbal activity that takes place between individuals, whereas language is the faculty, or the 'organ', that makes speech possible. Ever since Aristotle, language has been regarded as an activity that links *sound and meaning*, and requires therefore the coordination of two distinct systems: a phonetic system that receives and produces sounds (the *sensory-motor* component of language) and a cognitive system that gives meanings to sounds (the *semantic* component of language). Recently, however, it has been acknowledged that a third system must exist in order to perform an additional type of processing. This third component of the faculty of language is the system responsible for *syntax*, the set of rules that all combinations of words must follow to be accepted as valid linguistic expressions.

Chomsky has convincingly argued that syntax and semantics, although intimately interrelated, are distinct entities. He demonstrated this point with the classical sentence *"colorless green ideas sleep furiously"*, which is nonsense in terms of meaning and yet it is correct in terms of syntax. Most importantly, Chomsky recognized that it is this third component of language that is capable of generating an unlimited number of expressions from a finite set of elements. It is syntax, in other words, that is responsible for *recursion*, and for that reason one can rightly regard it as the generative engine, or the computational machine, of language.

The very special role that syntax plays in language has induced Hauser et al. (2002) to propose two distinct definitions of language. The faculty of language in the broad sense (FLB) is formed by all three systems mentioned above (phonetics, semantics and syntax), whereas the faculty of language in the narrow sense (FLN) contains syntax alone. The rationale of this proposal is that "*FLN is the only uniquely human component of the faculty of language*", whereas the other two come from our animal ancestors. According to this proposal, in short, all animals are capable of communication, but language exists only in our species because only humans have evolved the generative engine of syntax.

## 8.2 Sebeok's Definitions of Language

Thomas Sebeok (1920–2001) reached his conclusions on language by elaborating the concepts proposed by Jakob von Uexküll (1864–1944) and by Juri Lotman (1922–1993). According to Uexküll, every animal perceives the world with internal means and lives therefore in a subjective environment of its own making that he called *Umwelt*. At the same time, animals perceive their own body with a system that Uexküll called *Innenwelt*, and it is this *inner world* that is ultimately responsible for their behaviour (von Uexküll 1909).

Juri Lotman gave the name *semiosphere* to the world of culture, and regarded that term as the cultural equivalent of the name *biosphere* that is often used to describe the biological world. But Lotman was also a follower of Saussure, and inherited from him the idea that language is a concrete entity called *Parole* that is based on an abstract system called *Langue*. According to Saussure, Langue is the system that lies at the very heart of culture, and for that reason Lotman called it the *primary modelling system* of our species (Lotman 1991).

Thomas Sebeok accepted both the idea of a modelling system proposed by Lotman, and the concepts of Umwelt and Innenwelt proposed by Uexküll and argued that there is a deep relationship between them. More precisely, he concluded that Innenwelt and Umwelt together form a system that he described as the *primary modelling system* of animals. Von Uexküll, furthermore, had shown that most animals are capable of interpreting the world, and Sebeok proposed that this ability represents their *second modelling system*. Man inherited his primary and secondary modelling system from his animal ancestors, whereas language was a unique evolutionary development and represents therefore a *third modelling system* in our species.

It is worth noticing that the 'primary modelling system' of Sebeok is the whole set of brain-modelling faculties that we have inherited from animals, and this is slightly more general than the 'faculty of language in the broad sense' (FLB) defined by Chomsky. In the same way, Sebeok's 'third modelling system' includes all modelling faculties that evolved exclusively in our species and it is slightly more general than Chomsky's 'faculty of language in the narrow sense' (FLN). There are, in other words, only marginal differences between the definitions of language proposed by Chomsky and by Sebeok, and in this respect it is legitimate to say that they used different terminologies for largely similar purposes. What they profoundly disagreed about, instead, was the process that lies at the very heart of the language faculty.

## 8.3 The Bone of Contention

Noam Chomsky and Thomas Sebeok have been the architects of two major theoretical frameworks for the study of language and the founding fathers of two research fields that today are known respectively as *biolinguistics* and *biosemiotics*.

Chomsky's most seminal idea is the concept that our ability to learn a language is *innate*, that children are born with a mechanism that allows them to learn whatever language they happens to grow up with (Chomsky 1957, 1965, 1975, 1995, 2005). That inner mechanism has been given various names – first *Universal Grammar*, then *Language Acquisition Device (LAD)*, and finally *Faculty of Language* – but its basic features remain its *innateness* and its *robustness*. The mechanism must be innate because it allows children to master an extremely complex set of rules in a limited period of time, and it must be robust because language is acquired, generation after generation, in a precise sequence of developmental stages. For this reason, Chomsky concluded that the rules of grammar and syntax must be based on very general principles of economy and simplicity that are similar to the rules of the *Periodic Table* in chemistry or to the *Principle of Least Action* in physics (Baker 2001; Boeckx 2006).

Thomas Sebeok maintained that language is first and foremost a modelling system, the quintessential example of semiosis, and that 'interpretation' is its most distinctive feature (Sebeok 1963, 1972, 1991, 2001). He forcefully promoted the Peirce model of semiosis, which is explicitly based on interpretation, and insisted that semiosis is always an interpretive activity. Sebeok underlined that concept in countless occasions and in no uncertain terms: "*There can be no semiosis without interpretability, surely life's cardinal propensity*" (Sebeok 2001, p. 68).

This is the bone of contention between the two frameworks. Is the faculty of language a product of universal principles or of interpretive processes? Chomsky insisted that the ontogenesis of language must be precise, robust and reproducible like the ontogenesis of any other physiological faculty, and cannot therefore be left to the vagaries of interpretation. Sebeok insisted that language is semiosis and that semiosis is always an interpretive process, so it cannot be the result of mathematical principles or physical laws. This is what divides biolinguistics from biosemiotics. The concept that language is based on universal principles is not compatible with the idea that it is based on interpretation.

## 8.4 A Third Foundation of Language

There is ample evidence that animals are capable of interpreting the world, and this clearly means that *interpretive semiosis* is a reality. But it is also evident that the rules of the genetic code do not depend on interpretation because they have been the same in all living creatures and in all environments ever since the origin of life. The logical conclusion is that there are two distinct types of semiosis are in Nature. *Organic semiosis* appeared with the genetic code and was responsible for the origin of many other organic codes in the 3 billion years of cellular evolution. *Interpretive semiosis* appeared later and evolved exclusively in animals because only nervous systems can produce it (Barbieri 2011).

This distinction between coding and interpretation suggests that there is, in principle, a third foundation of language. More precisely, it suggests that the generative mechanism of language is provided neither by universal laws nor by interpretation but by biological codes. As soon as we take codes into consideration, in fact, we have a natural alternative to the opposite claims of biolinguistics and biosemiotics.

(a) The major claim of Biolinguistics is that the development of language cannot be left to the vagaries of interpretation and must be precise, robust and reproducible like the development of any other faculty. This goal, however, does not necessarily require universal laws. The ontogeny of language would be precise, robust and reproducible even if it is based on biological codes rather than universal principles. The genetic code, for example, guarantees precise, robust and reproducible features in all living system, and has properties that are even more universal than those of universal grammar. Language does indeed require rules, but the rules of grammar or the principles and parameters of syntax are much more likely to be the result of coding rules rather than the expression of universal phenomena like the Periodic Table or the Principle of Least Action. This is because the rules of life are produced by living systems and evolve with them, whereas the rules of mathematics and physics are not subject to historical change.

(b) The major claim of Biosemiotics is that the faculty of language is based on semiosis, and this is not compatible with the idea that the rules of language come from universal laws. This however does not mean that the ontogeny of language is necessarily an interpretive process because organic semiosis exists and it is based on coding not on interpretation. The ontogeny of language, in other words, could well be a semiotic process based on codes and in this case it would not be subject to the vagaries of interpretation.

The idea that language is founded on biological codes, in conclusion, removes the two obstacles that so far have divided biolinguistics from biosemiotics, and at the same time it gives us a new theoretical approach to language. It suggests that there have been biological codes at the origin of language just as there has been a genetic code at the origin of life and a neural code at the origin of mind. Before dealing with this issue, however, we must discuss the biological milieu that provided the ground for the origin of language.

## 8.5 A Juvenile Ape

In 1926, Luis Bolk, professor of anatomy at Amsterdam university, proposed the *fetalization theory*, the idea that the origin of man was due to the extension of foetal or juvenile features to the adult phases of life (Bolk 1926). The idea was not new (Geoffroy Saint-Hilaire had mentioned it in 1836), and the phenomenon had been described in many other species with names such as *paedogenesis* (von Baer 1866), *neoteny* (Kollmann 1885), and *paedomorphosis* (Garstang 1922). But it was Luis Bolk who turned that idea into a compelling doctrine by the sheer number of data with which he supported it.

In all primates, the foetus and the newborn child have big brains (in respect to body size), flat faces, thin brow-ridges, small teeth and jaw, light skin and sparse body hair, but only man retains all these features in adult life. Similarly, the front-to-back axis of the head is perpendicular to the axis of the trunk in the foetus and in the newborn child of all primates but only man maintains that angle of the cervical flexure throughout life, and that is what allows him to have a horizontal line of sight while standing erect, whereas all other primates can look around while walking on all fours (Bolk 1926).

Big brains, flat faces, reduced body hair and upright posture are unmistakable marks of humanity, and are undeniably present in the foetal stages of all primates, so there is little doubt that an extension of these juvenile features, i.e., a process of fetalization, did take place in our ancestors.

It is also known that environmental changes can affect development and produce either a retardation or an acceleration of sexual maturity in respect to body growth. In axolot, for example, an abundance of water in the environment favours an extension of the larval stage, whereas a dry climate induces a quick metamorphosis to adult stage (Kollmann 1885; Gould 1977). It is likely, therefore, that changes in the environment provided the initial pressure for changes in the embryonic development of our ancestors and we know that vast climatic changes did take place in Africa in the past 10 million years.

We conclude that the fetalization theory provides a sound theoretical framework for the origin of man. At the same time, however, we must not forget that neoteny and many other processes of *heterochrony* (changes in timing) have taken place in countless animal species, but have never produced a cognitive system based on symbols. Our problem, therefore, is to find out what was it that made the difference in our species. More precisely, among all the evolutionary processes that shaped the human body, we are looking for those that created the preconditions for the origin of language.

## 8.6 Fetalization and Brain Wiring

In the 1940s, Adolf Portmann calculated that our species should have a gestation period of 21 months in order to complete the processes of foetal development that occur in all other mammals (Portmann 1941, 1945). A newborn human baby, in other words, is in fact a *premature foetus*, and the whole first year of his life is but a continuation of the foetal stage.

This peculiarity of human development is due to the fact that fetalization leads to an extended foetal period and therefore to a greater foetus at birth, but this process is severely constrained because the birth canal can cope only with a limited increase of foetal size. During the evolution of our species, therefore, any extension of the foetal period had to be accompanied by an anticipation of the time of birth. The result is that the foetal development of our species became divided into two distinct phases – intrauterine and extra-uterine – and eventually the extra-uterine phase came to be the longest of the two.

It is not clear why this evolutionary result is uniquely human, but it is a historical fact that it took place only in our species. In all other mammals, foetal development is completed *in utero*, and what is born is no longer a foetus but a fully developed infant that can already cope with the environment.

The crucial point is that the last part of foetal development is the phase when most synaptic connections are formed. It is a phase of intense *brain wiring*. The fetalization of the human body has produced therefore a truly unique situation in our species. In all other mammals the wiring of the brain takes place almost completely in the dark and protected environment of the uterus, whereas in our species it takes place predominantly outside the uterus, where the body is exposed to the lights, the sounds and the smells of a constantly changing environment. In our species, in short, the difference between intrauterine and extra-uterine foetal development created the conditions for *two distinct types of brain wiring*, and this did have far reaching consequences.

The brain wiring that occurs in the last phase of foetal development provides the neurological basis for the mental models that the organism is going to use throughout its life. If that phase occurs in the highly stable and reproducible environment of the uterus, the operations of brain wiring follow a pre-established sequence of steps and generate a cognitive system that has been highly conserved in evolution. In our species, instead, the last phases of fetal development have been progressively displaced outside the uterus, in a radically different environment, and that created the opportunity for a radically new experiment in brain wiring. That was the precondition for the evolution of a uniquely human cognitive system, but let us not forget that a precondition for language was not yet language. It was only a potential, a starting point.

## 8.7 Brain Size and Language Genes

The human brain is about three times larger than the brain of any other primate, even when body weight is taken into account. This means that the first and second cognitive systems that we have inherited from our animal ancestors required, at most, a third of our present brain size. The other two thirds could be explained, in principle, by a further extension of our animal faculties, but this is not what happened. We have not developed a sharper eyesight, a more sensitive olfactory system, a more powerful muscular apparatus, and so on. As a matter of fact, our physiological faculties are in general less advanced than those of our animal relatives, so it was not their improvement that explains our increased brain volume. It is possible therefore that the brain increase that took place in our species was largely due to the development of those new faculties that collectively make up our *third* cognitive system, the system that eventually gave origin to language. It is also possible, on the other hand, that an increased brain size was only a *precondition* for language, not a result of it, and what caused it therefore was some other biological process.

Another major issue concerning the origin of language is the role that genes played in it. Today there is still some disagreement on the definition of the genes of language, but it seems reasonable to say that they are genes whose mutations produce heritable changes in the faculty of language. The point is that there are many examples of such genes, but the outstanding conclusion that has come out of their study, so far, is that virtually all of them are also present in animals. All known genes of language, in other words, are genes of the primary and secondary modelling system that we have inherited from our animal ancestors. Future discoveries may modify this conclusion, of course, but not much. The reason is that we share about 98 % of our genes with the chimpanzees, so the number of uniquely human genes is bound to be small (Sibley and Ahlquist 1984).

The experimental evidence, in short, tells us that the genes of language do exist, but that virtually all of them also exist in animals and this means that our third cognitive system was largely built with animal genes, not with uniquely human genes. This in turn implies that language was largely the result of epigenetic processes that were based on animal genes and produced a uniquely human result. But can we account for such a result? Can we explain how a new complex cognitive system can be produced almost entirely by epigenetic processes? The answer, surprisingly, is yes. There is indeed a model based on a classic experiment in embryology that gives us an illuminating guideline.

## 8.8 The *Cerebra Bifida* Model

In vertebrate embryonic development, the heart arises from two primordia that appear on the right and left side of the developing gut, and then migrate toward the centre of the body and fuse together in a single organ. If fusion of the two primordia is prevented by inserting an obstacle between them, each half undergoes a spectacular reorganization and forms a complete and fully functional beating heart (Fig. 8.1). The formation of the two hearts, furthermore, is followed by the development of two fully functional circulatory systems, and the animal goes through all stages of life in a double-heart condition that is known as *cardia bifida* (DeHaan 1959).

This classic experiment shows that two profoundly different bodies, one with a single heart and the other with two hearts, can be generated *without any genetic change at all.* A modification of the epigenetic conditions of embryonic development is clearly an extremely powerful tool of change, and may well be the key to human evolution.

The gradual extension of our foetal period together with the constraint of the birth canal have split the foetal development of our brain into two distinct processes, one within and one without the uterus, whereas in all other mammals it has remained a single process that takes place entirely within the uterus. This splitting of our brain development is a condition that can be referred to as *cerebra bifida*, in some

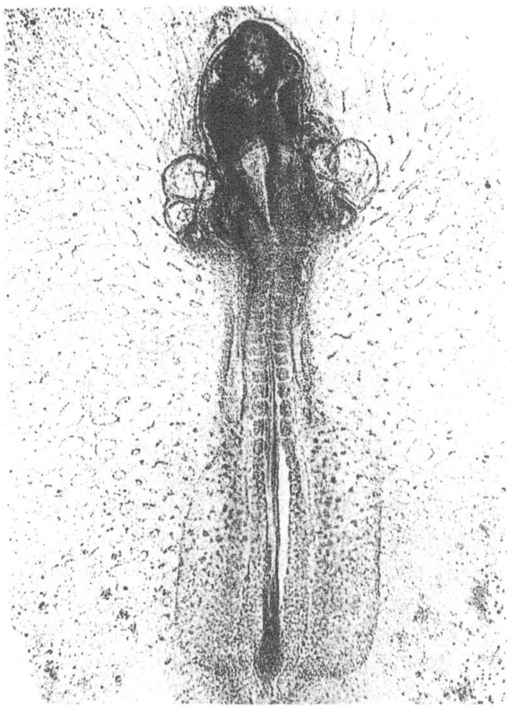

**Fig. 8.1** The heart arises from left and right cardiac primordia that move together and fuse in the midline. If fusion is prevented each half forms a complete and fully functional heart, as seen in this 2-day-old chick embryo, a condition that is known as *Cardia bifida* (DeHaan 1959)

ways analogous to *cardia bifida*, except that in the case of the heart the two organs arise from a separation *in space* whereas in *cerebra bifida* they are produced by a separation *in time* between intrauterine and extrauterine development.

The *cardia bifida* experiment is illuminating because it shows that an organ can literally be duplicated by a separation of its developing tissue into two compartments, and it is possible therefore that the enormous increase in brain size that took place in human evolution was largely due to the separation of the foetal development of our brain into two distinct phases, a conclusion that can be referred to as 'the *cerebra bifida* model' on the evolution of man.

A second illuminating result of the *cardia bifida* experiment is that no new genes are required for the duplication of the organ, and this may explain why new genes were not strictly necessary for building a new cognitive system in our species. The same set of genes that we have inherited from our animal ancestors could well have produced a different cognitive system simply by operating in extrauterine conditions that induced a different type of brain wiring.

It must be underlined, however, that an increase in brain size accounts only for the hardware of a new cognitive system, not for its software. In order to understand the origin of language, in other words, we need to uncover the actual processes that created our specifically human characteristics.

## 8.9 Mind and Reality

The discovery that perceptions are produced by the brain implies that we live in a world of our own making, and this leads to the idea that there is an unbridgeable gap between mind and reality. But how does our brain construct our perceptions? At the end of the eighteenth century, Immanuel Kant argued that the signals carried by our senses become meaningful to us only when they are interpreted in terms of abstract categories – in particular space, time, object and causality – which are not derived from experience because they are *a precondition* for acquiring experience. Kant (1781) concluded that these categories are *a priori*, because they are innate, and are *transcendental* because they transcend, or come before, experience. According to Kant, in other words, our knowledge of the world relies on two unknowable entities: we have no access not only to the raw data of reality (the *things-in-themselves*) but also to the *a priori* categories that allow us to build mental objects and experience the world. In such a situation, what kind of knowledge do we have of reality? And where do the *a priori* categories of knowledge come from?

An answer to these questions came from Darwin, and it is the idea that mental objects are *adaptations* that allow us to cope with the world. Natural selection allows organisms to become increasingly adapted to the environment, and can be regarded therefore as a process that reduces the distance from reality, that allows animals to catch increasing amounts of reality. This is because mental images reflect *relationships* between things, and are specifically selected so that the relationships between them represent at least some of the relationships that exist between objects in the physical world.

Our perceptions, in short, must reflect some aspects of reality, otherwise we would not be able to survive. François Jacob has expressed this concept with admirable clarity: "*If the image that a bird gets of the insects it needs to feed its progeny does not reflect at least some aspects of reality, there are no more progeny. If the representation that a monkey builds of the branch it wants to leap to has nothing to do with reality, then there is no more monkey. And if this did not apply to ourselves, we would not be here to discuss this point*" (Jacob 1982).

This idea, furthermore, does not apply only to the perceptions of the physical world but also to animal behaviour. In the 1940s, Konrad Lorenz discovered that in many species of birds the newly hatched chicks follow the first object they see moving in front of them and take it as their natural mother (Lorenz 1971). This *following response* is a form of learning because the chicks acquire a specific behaviour, but it is a learning developed by the species in evolutionary times. Lorenz called it *phylogenetic learning*, and sharply distinguished it from the *ontogenetic learning* that takes place during the development of an individual.

More precisely, Lorenz argued that there are indeed mental categories that are present at birth and are therefore *innate*, or *a priori*, but underlined that what is *a priori* for the individual, is *a posteriori* for the species. In the case of the individual, at any rate, space, time, object and causality appear to be innate, or *a priori*, categories, as Kant had maintained, but Jean Piaget showed that this is not exactly the case.

## 8.10 Steps in the Ontogeny of Mind

The classical studies of Jean Piaget on the cognitive development of the child have shown that our mental categories (in particular the concepts of object, space, time, causality and number) arise step by step throughout infancy in a sequence of four distinct periods (sensorimotor, preoperational, operational and formal) each of which is subdivided into stages (Piaget 1954, 1960).

1. In the first period (from birth to 2 years of age), infants construct the concepts of object, space, time and causality in parallel with the development of the hand-eye coordination. During the first month of life, infants can move their arms but do not realize that the hand they see is their own. By the age of 6 months they have the ability to follow moving objects and to remember them when they go out of sight. When an object disappears, however, infants expect to find it not where it went out of view but where it was first seen, as if the object had gone back to the place where it belongs. Only a few months later infants acquire the notion that objects perform continuous movements in space and time. As for the concept of causality, it starts with the belief that events are caused by willing them into existence and gradually changes when the child abandons the link between wishing and causality, and realizes that most connections between events are physical rather than mental.
2. During the second period (from 2 to 5 years of age), children begin to reason from memory and analogy, perform imitations and start using symbols, with objects represented symbolically either by words or by other objects. These first symbols are *private* tools that children use to think aloud and to internalize thoughts, not to communicate with other people, in sharp contrast with the vocalizations of animals which are used for *public* reasons and have nothing to do with the internalization of thoughts. Most utterances in kindergartens, for example, are monologues by which children comment their own feelings and actions. True social communication appears only at a later stage.
3. During the third period (from 5 to 10 years of age), children acquire the ability to perform mental operations on objects that are concretely present. They can construct hierarchies and begin to develop concepts such as number, weight and volume. Social communication also starts developing in this period and children become interested in exchanging information with other people.
4. In the fourth period (from 10 to 14 years of age) children start conceiving alternative scenarios to any given situation and begin to elaborate conjectures that could be true or false. Propositional thinking is possible and the mind becomes actively engaged in the planning of future events.

Piaget underlined that the age at which children reach these stages can vary from one individual to the other, but the sequence of the stages is invariant. This means that the beginning of one stage presupposes the completion of all preceding stages, as it is the case in the ontogeny of all organs in the body.

## 8.11 Interacting Brains

Children are born with the *potential* to learn any language they grow up with, and the studies of Jean Piaget have suggested that this ability is acquired in stages during the first years of life. We have therefore the problem of understanding how this development takes place, and luckily we do have a few interesting clues. Nancy Soja, Susan Carey and Elizabeth Spelke, for example, have shown that two-and-a-half-year old children can literally invent new rules about imaginary names (Soja et al. 1991).

A T-shaped metal object (a hydraulic joint) was given the made-up name *blinket*, while a piece of dough was called *dax*, and the children were invited to find other *blinket-objects* and other *dax-objects* in the room. It turned out that they identified as 'blinket' any T-shaped object, even if made of wood, cardboard or plastic. The name 'dax', instead, was given only to dough-made objects, irrespective of their shape, and was never used for pieces of wax or jelly, even when these had the same shape and size of the original piece. Clearly the children deduced that 'blinket' means a particular shape, not a material substance, while 'dax' was referring to a material substance, not to a shape. This shows that by the age of two-and-a-half-years children are not passive learners but are actively *making hypotheses* about the meaning of words.

The long-term acquisition of new neural functions, on the other hand, requires a permanent change in brain-wiring and this is achieved by the formation of new synapses in critical stages of development. But is this true also in the case of language? Are there *critical* stages in language development as there are in the development of all the other faculties of the body? The answer to this question has come from the study of children who have been raised in the wild by animals (the so-called *wolf children*, or *feral children*). Their physiological functions are perfectly normal, but their potential to learn a language is severely compromised and some linguistic properties are irreversibly lost if they are not developed within some crucial periods of time (Maslon 1972; Shattuck 1981).

The evidence from feral children tells us two important things. One is that critical stages do exist in the development of language; the other is that language is crucially dependent upon *human interactions*. It is not like the *following response* in newly-hatched chicks that can be activated by virtually anything.

In the ontogeny of language, in short, there are two complementary faces. The existence of critical stages is common to all animals, whereas the existence of critical stages that depend upon human interactions is a uniquely human phenomenon.

Another important clue comes from the studies of the 'creole' languages, and is the fact that the major role in the making of new linguistic rules appears to be played by children (Bickerton 1981). This makes us realize that the development of language crucially depends on interactions that take place first between child and mother and then between child and other children. What brings language into existence, in other words, is not the brain on its own but *a community of interacting people*. This conclusion is a version of what has become known as *Distributed Language*, the idea that language is not inside the individual head but is 'distributed' in a population (Cowley 2011).

## 8.12 A Hidden Asymmetry

The study of the faculty of vision has revealed that the formation of optical images takes place in stages and in different areas of the brain. The incoming signals are first processed on the retina, then sent via the optic nerve to the midbrain, and after the optic chiasm (where 50 % swap direction) are transmitted to the visual cortex, at the back of the head, where they are further processed by groups of cortical cells arranged in distinct areas. This *distributed* structure made of many neural processing centres has been found in all other faculties of the body, and what is particular important to our purposes is that language is no exception.

The cortical areas discovered by Broca and Wernicke, for example, are involved, in different ways, in the production of speech, whereas another speech-related area, the *angular gyrus*, is concerned not with the production but with the decoding of speech. At the same time, it must be underlined that the speech production areas have an asymmetric position in the human brain, because they are present on one side only (usually in the left hemisphere) whereas most functions of the cerebral cortex are distributed symmetrically over the two hemispheres. This may not look surprising, at first, because there are other asymmetries in animal bodies, but it turned out that the speech-areas lateralization has a deep hidden meaning.

This was brought to light by Roger Sperry's studies of patients affected by epilepsy who had undergone the *split-brain* operation, i.e., the cut of the *corpus callosum*, the massive strand of nerve fibers that connect the two hemispheres (Sperry 1968, 1982). This chirurgical cut was effective in relieving the worse aspects of epilepsy, and apparently it allowed patients to lead a normal life. But something in their brain had deeply changed.

When the patients received signals that were designed to reach only one hemisphere, it turned out that the other hemisphere had no knowledge of what had happened. When the corpus callosum is cut, in other words, the left hemisphere, which contains the speech centres, has no idea of what is being presented to the right. A patient may have an emotional response, for example a smile, when seeing a picture with the right brain, but if asked *why* he or she smiles, the verbalizing left brain cannot answer it.

These great discoveries show that our two hemispheres are so well integrated that we seem to have only one mind, whereas in fact we have *two* different minds that talk to each other via the *corpum callosum* and then talk to the outside world with one voice, controlled by the left mind. During the evolution of our brain, in other words, there has been a splitting event that has committed our two hemispheres to two very different functions. The right hemisphere has taken on the emotional processing while the left hemisphere has been put in charge of the verbalizing activity.

It must be underlined that the two symmetrical hemispheres normally perform in slightly different ways. The left and the right eye, for example, receive signals from slightly different angles and this allows the two hemispheres to construct a three-dimensional perspective. In this case, however, the two hemispheres are performing the *same* function from two different points of view, whereas in the case of language the two hemispheres perform two *completely different* functions.

We conclude that there are two very different aspects in language: on the one hand it resembles the faculties of our animal ancestors because it is *distributed* throughout the brain; on the other hand, it is a specifically human faculty because it is based on a uniquely human asymmetry that has split our cerebral cortex into two halves that have profoundly different functions.

## 8.13 Rules of Neural Development

Beyond a certain level of complexity, the genes cannot specify all details in a developing body, and Nature has resorted to a new strategy: the genes determine only a few basic structures, and all remaining features are left to the encounters with environmental processes that in normal conditions take place with a high degree of regularity.

In the case of the *following response* discovered by Konrad Lorenz in newly hatched chicks, for example, the neural circuit of the response is already in place at birth, and only needs to be activated by the encounter with a moving object. Ideally this is the mother bird, but the genes cannot provide a detailed description of her and Nature has invented a far simpler activating event: the appearance in the visual field of anything that moves. This is a far from perfect solution, but in a natural context it is highly likely that the first moving object that appears before the chicks is the mother bird, and in practice this is good enough to guarantee the survival of the species.

It will be noticed that the instruction "*follow the first moving object*" is not a physical law, a principle of logic or a process of interpretation. It is, much more simply, the rule of a code. It is a rule that creates a link between a visual event in the retina and a motor circuit in the brain, just as a rule of the genetic code, for example, creates a link between a triplet of nucleotides and an amino acid. In both cases, the rules of a code are physically implemented by structures called *adaptors*, and it is precisely the presence of these structures that reveals the existence of a code. In the genetic code the adaptors are the transfer-RNAs, whereas in the code of the *following response* the adaptors are the neural circuits that establish a connection between the retina of the newly hatched chicks and the neurons of their walking system.

We find the same logic in countless other examples of embryonic development. The building of the nervous system, for example, takes place with a strategy that consists in an overproduction of neurons followed by the programmed suicide of those that do not make the right encounters. The point is that what constitutes a '*right encounter*' varies from place to place and is determined by coding rules of the type '*in the absence of a specific nerve growth factor, activate the suicide genes*'.

Another example of neural rules comes from the wiring of the nervous system, a process that is based on the making and breaking of synaptic connections. Initially the system is over-connected because neurons generate an excess number of synapses. These, on the other hand, are continuously broken and reformed, and only the synapses that are repeatedly reconnected become stable structures. In this case, the coding rule is '*in the absence of reinforcement eliminate the synapse*'. This mechanism continues to operate long after birth and in some part of the brain it goes on indefinitely, thus providing the means to form new neural connections throughout the life of an individual.

## 8.14 Specifically Human

Animals receive signals from the environment, transform them into mental images and perform mental operations that allow them to *interpret* what goes on in the world. Interpretation, in turn, is an act of *semiosis* because it is based on signs, not on physical necessity. We have learned from Peirce that there are three major types of signs in Nature – *icons*, *indexes* and *symbols* – and this gives us the problem of understanding the use that animals make of them.

The mental representations of the brain have a surviving value when they catch at least some aspects of reality, and to this purpose natural selection can definitely use the relationships that are based on icons and indexes, because they reflect properties that effectively exist in Nature.

Icons, for example, are signs that represent objects in virtue of a *similarity* that exists between them, and similarities are useful because they allow animals to recognize objects from a few general features.

Indexes are signs that represent objects because of a *physical link* between them. Footprints, for example, are signs of preceding animals, smoke is a sign of fire, and so on. Indexes are the basic tools of learning because they allow animals to infer the existence of something from a few physical traces of something else.

In the case of *symbols*, however, the situation is totally different. A sign is a symbol when its association with an object is created by an entirely *arbitrary* link. There is no similarity and no physical link between a name and a person, for example, or between a flag and a country. Symbols allow us to make arbitrary associations and build mental images of hypothetical events (projects), of abstracts things (numbers), and even of non-existing things (unicorns).

The arbitrariness of symbols means that natural selection *cannot* favour them to represent the real world because arbitrary relationships would increase rather than decrease the distance from reality. Natural selection, in short, is actively working *against* the use of symbols as a means to represent the physical world.

This explains why animals have *not* evolved language, but it gives us the problem of understanding why we did. A massive and systematic use of symbols is indeed a specifically human activity, but how do we account for it? It is an experimental fact that children make an active use of symbols at very early stages of development, but how could that ability have evolved? What good is to a child the ability to make arbitrary connections, to believe in non-existing objects?

## 8.15 The Third Cognitive System

We have inherited two cognitive systems from our animal ancestors. The first is the ability to perceive the world, the second is the ability to interpret what goes on in the world with processes of extrapolation based on icons and indexes.

When a human baby is born, however, he is thrown into a situation that no animal has ever experienced. He is expelled from the uterus when his foetal development is barely half-way through, and the two animal cognitive systems cannot help because they are not yet fully developed. In such a situation a newborn baby has no choice but resorting to something else, and that is why it starts building a *third* cognitive system. To this purpose, all a human baby can do is uttering sounds, and that is precisely what the third cognitive system, i.e., the faculty of language, is about: *the use of sounds to attract attention*.

At first sight it may seem that the enterprise is doomed to fail. A human baby is like a pilot who is ejected from its spaceship before reaching destination, and the chances of survival seem next to nil. In our case, however, we must take into account that the extrauterine phase of human development was very short at the beginning of our evolutionary history, and probably it took a few million years to reach the present value (it would be interesting to find out how long was that phase in the *Australopithecines* and in *Homo erectus*, for example). The faculty of language, in other words, has evolved in stages during the history of our species, and it is developed in stages during the ontogeny of every human being.

At birth, a human baby can barely move its body and has a blurred image of what is around him, so an innate response, like '*follow whatever is moving in front of you*', would not even make sense in our species. A newborn human baby can only emit sounds, and soon it discovers that crying, for example, is an effective way of getting attention. Its brain, on the other hand, is making synaptic connections all the time and the new neural circuits establish real connections between *wishing* something and *getting* it. This is how the brain of a child starts believing that events are caused *by wishing them into existence*. This is an arbitrary assumption, of course, and in later stages the child will have to abandon it, but in the first few months of life it is an effective means of exploiting its scarce resources.

Another step further takes place when the baby pays attention to what it is hearing, and starts making connections between emitting and receiving sounds. This sets in motion what Cowley (2007) has described as a dance of back-and-forth interactions where mother and child are continually regulating and adapting to each other's activities. Eventually, the child learns that by *mimicking* the adult's utterances he or she gets increased attention, and starts emitting vocal patterns that adults *interpret* as speech. “By their first birthday”, says Cowley, “children regulate their activity by using *what adults call* words and gestures.”

In the course of her developing relationship with the child, the mother is freely using the symbolic expressions she is used to, and that predisposes the child to take in those expressions as a normal part of life. The important point is that the new neural connections that the child is making all the time give *a neurological basis* to the symbolic activity of the brain, and this is why such activity not only survives but takes on increasingly complex new forms. This is probably what made us human. The ability, or perhaps the *necessity*, to make arbitrary mental connections and build a world of symbolic objects.

## 8.16 Our Rational Faculties

The origin of man has been associated with a whole constellation of mental and physical features, but it has been reported that most of them are also present, in varying degrees, in nonhuman animals. This is true also for ‘higher’ faculties, such as the ability to count, to form abstract concepts and to anticipate future events, that for a long time have been traditionally regarded as specifically human.

An octopus, for example, can distinguish between a triangle and a square, or between a vertical and a horizontal rectangle (Sutherland 1964). Monkeys can represent numbers up to nine (Brannon and Terrace 2000), and in some cases up to thirty (Cantlon and Brannon 2006). Some birds can count from one to six (Pepperberg and Carey 2012). Bats have the ability to form three-dimensional models of their environment and exploit them in their hunting of insects (Yartsev and Ulanovsky 2013). Rats that are maneuvering a maze can recall past events and make plans for the future, on the underlying assumption that the maze has not changed (Corballis 2013; Suddendorf 2013). Whales learn from encounters with predators to develop new routes for their future migrations (Alerstram et al. 2003). Most importantly, it has been shown that animals have the ability to attribute beliefs, desires, and intentions to other animals, that is, they have a *theory of mind*. Jays, for example, have the habit to hide food, but if they are being watched they will later return and relocate it (Daily et al. 2010).

There is ample evidence, in short, that animals can perform complex abstract operations and this means that many of our mental faculties have evolved from our animal ancestors. It must be underlined, however, that the animal ability to interpret the world is based exclusively on icons and indexes. Even a theory of mind is accounted for by these two types of signs only and does not require symbols. What we have inherited from our animal ancestors, in other worlds, is the faculties that are based on icons and indexes, not the one faculty that requires an extensive use of symbols.

The interesting point is that many mental processes that we traditionally regard as uniquely human, such as counting, anticipating the future and predicting other people's behaviour, are entirely based on icons and indexes, and it is for this reason that – in rudimentary forms – they are also present in animals. What made us human, in other words, is not our *rational* faculties.

According to Aristotle, man is endowed with three 'souls' (*anima vegetativa*, *anima sensitiva*, and *anima rationalis*), whereas animals have the first two souls and plants only the first one. In this classical scheme, only man is blessed with *rationality*, a conclusion that has been taken for granted from antiquity up until very recently. Now we are beginning to realize that animals too have the basic elements of *anima rationalis*, and what is uniquely human is another type of 'soul', a type that can be referred to as *anima irrationalis* because it is the ability to make *irrational assumptions*, to take symbols for real and live in an imagined world.

## 8.17 The Codes of Language

One of the major theories of language is the idea, first expressed by Saussure, that *language is a code*, a set of rules that give meanings to sounds. This view is often referred to as *the code view of language*, because it conveys the idea that language is a mapping, a software that is independent from the hardware of the brain and consists in a set of *disembodied* coding rules.

Against this view there has been a whole constellation of counterproposals which are all based on the idea that language is physiology, behavior and sensory-motor activity, in short, a fully *embodied* faculty (Harris 1981; Maturana and Varela 1987; Donald 1991; Love 2004).

The underlying assumption of these proposals is that no physiological faculty can be based on coding rules, and yet there is an outstanding example that this is precisely what takes place at the very heart of life. The apparatus of protein synthesis is the physiological system that manufactures proteins according to the rules of the genetic code and there is no way that this code can be reduced to a set of disembodied rules. More than that. Only a process that employs a code has the potential to produce genuine novelties in Nature, because only a code can establish *arbitrary* links between independent objects.

Protein synthesis is indeed a paradigmatic example of a coded biological process, and this is why it can be taken as a model for other such processes. It is possible, in particular, to establish a parallel between the apparatus of protein synthesis and the intermediate brain, where one is manufacturing proteins and the other is manufacturing mental states. This gives us a *biological code view of language* that is profoundly different from the Saussurean view because the codes of language are no longer abstract rules but physiological components of the language faculty.

But what exactly are the codes of language? This is an issue where again it is helpful to keep in mind the model of protein synthesis. The genetic code is implemented by a set of molecular adaptors that physically provide the means for manufacturing proteins, and proteins, in turn, provide the means for assembling supramolecular structures into a living cell. The genetic code, in other words, is the starting point of a long chain of events that produce not just proteins, but an integrated supramolecular system.

In the case of language, we have seen that a human baby starts from mimicking the sound patterns of the adults, then goes on to making an increasing use of symbolic expressions and by the age of two-and-a-half has fully acquired the ability to invent new symbolic associations between names and objects. But these are still individual pieces of symbolic activity, the equivalent of isolated proteins that are not yet producing a whole cell. When it is then that the pieces come together into a coherent symbolic system? This takes place when children become capable of arranging words into stories that *make sense* to their community, and we reach in this way the conclusion that the codes of language are in fact *the codes of storytelling*, the rules that make that it possible to tell countless stories about countless imaginary worlds in countless different ways.

What is particularly worth noticing is that this idea brings the origin of language in line with the other great events of macroevolution, like the origin of life and the origin of mind. The common factor is that they were all associated with the appearance of new manufacturing systems based on new biological codes.

# Part III
# A New Science of Life

# Chapter 9
# Biosemiotics

Biosemiotics is the idea that semiosis (the production of signs) is such a fundamental component of life that one cannot exist without the other. But where does semiosis come from? On this point, biosemiotics is split into two opposite camps that represent its two historical starting points.

The synthesis of biology and semiotics that is known as *biosemiotics* was developed independently in two fields that lie at the opposite ends of academia. The first origin took place in molecular biology as a result of the discovery of the genetic code (the name *molecular biosemiotics* was coined by Marcel Florkin, in 1974, precisely to designate the study of semiosis at the molecular level). The second origin took place in the humanities and was masterminded by Thomas Sebeok in two distinct stages. In 1963, Sebeok extended semiosis from human culture to animals and founded the new research field of *zoosemiotics* (Sebeok 1963). More than 20 years later, he made a second extension from animals to all living creatures and called it *biosemiotics* (Anderson et al. 1984; Sebeok and Umiker-Sebeok 1992; Sebeok 2001).

Other proposals have appeared in between and have given origin to various *schools* of biosemiotics, but the original dichotomy about the nature of semiosis has remained.

In biology, the presence of semiosis at the molecular level is documented by the existence of the genetic code in all cells, and this implies that *molecular (or organic) semiosis is produced by coding*.

In the humanities, the dominant view is the Peircean concept that semiosis is always an interpretive process, and this implies that *semiosis is produced by interpretation*.

We have therefore two types of semiosis, one based on coding and one based on interpretation, and each of them represents phenomena that undoubtedly exist in Nature. There is ample evidence that animals are capable of interpreting the world, and this clearly means that *Peircean* (or *interpretive*) *semiosis* is a reality. But it

© Springer International Publishing Switzerland 2015

M. Barbieri, *Code Biology*, DOI 10.1007/978-3-319-14535-8_9

is also evident that the rules of the genetic code do not depend on interpretation because they have been the same in all living creatures and in all environments ever since the origin of life.

The logical conclusion is that both types of semiosis are present in Nature and represent two distinct evolutionary developments. This, however, is precisely what the Peirce followers do not accept. If there is mind, interpretation and semiosis in animals, they claim, it is because there have been forms of mind, interpretation and semiosis in all previous living systems, including the first cells.

Today this is the majority view in the field, and biosemiotics is almost universally identified with *Peircean biosemiotics*, a doctrine that is firmly based on Peirce's philosophy. At the same time, however, the line of research that started with the discovery of the genetic code and advocated a *scientific* approach to the codes of life has not been abandoned, and today it has become an independent field of research with the name of *Code Biology*.

What happened in the past, on the other hand, is still the best way to understand the present, and that is why this chapter is dedicated to a brief historical overview of biosemiotics.

## 9.1 Matter Controlled by Symbols

The discovery of the genetic code took place between 1961 and 1966 (Nirenberg and Matthaei 1961; Speyer et al. 1963; Niremberg et al. 1966; Khorana et al. 1966), and immediately inspired the idea of a deep parallel between the genetic code and the codes of language. This idea was vividly expressed by George and Muriel Beadle in 1966: "*the deciphering of the genetic code has revealed our possession of a language much older than hieroglyphics, a language as old as life itself, a language that is the most living language of all – even if its letters are invisible and its words are buried in the cells of our bodies*" (Beadle and Beadle 1966). But was this only a poetic metaphor or can we really say that the genetic code is the expression of a true molecular language?

A language is based on signs and can exist only in systems that make use of signs, i.e., in semiotic systems. The genetic code would be therefore the expression of a real molecular language only if the cell is a real semiotic system, i.e., only if signs exist inside the cell and are instrumental to its functions. This is the great potential implication of the discovery of genetic code, but can we prove it?

The idea that the cell is controlled by *symbols* was explicitly proposed for the first time by Howard Pattee at the symposia on theoretical biology organized by Conrad Waddington between 1966 and 1970, and was published in the proceedings of those symposia (Pattee 1968, 1972). The experimental evidence of the genetic code did not seem enough, on its own, to conclude that the cell is a semiotic system, but Pattee argued that it becomes enough when we combine it with the theory of self-reproducing automata developed by John von Neumann in the 1950s (von Neumann 1951, 1956, 1966).

Von Neumann showed that a self-reproducing system must necessarily contain a description of itself, and such a description must be categorically different from the construction it controls. Self-reproduction is a two-step process: the first consists in transferring the description to the descendants, and the second in using it to reconstitute the original system in each descendant. The description of a system, on the other hand, must be qualitatively different from the system ("*the map is not the territory*") so it is necessarily a set of entities that represent, or 'stand for', the material components, and therefore function as symbols. According to von Neumann, in short, a self-replicating system must be *a physical system controlled by symbols*.

The discovery that genes carry the information for the synthesis of proteins proved that the cell is a system that contains two distinct categories, a software and a hardware (a genotype and a phenotype). But the cell is also a self-reproducing system and Pattee concluded, on the basis of von Neumann's logic, that the genotype must be a *symbolic* description of the cell. This is the argument that he used to conclude that "*life is matter controlled by symbols*", a theme that he developed for nearly 40 years in various publications (Pattee 1968, 1972, 2001, 2008).

The idea that symbols exist at the cellular level was the first argument in favour of biological semiosis, and Marcel Florkin (1974) gave to the study of this process the name of *molecular biosemiotics*.

## 9.2 Physical Biosemiotics

A substantial contribution to the birth of molecular biology came, in the 1940s and 50s, from physicists who were fascinated by the idea that the study of life would disclose '*new laws of physics*', a prophecy made by two founding fathers of quantum theory, Niels Bohr (1933) and Erwin Schrödinger (1944). By the 1960s, however, any interest in new laws of physics had virtually disappeared, and most biologists acknowledged that life can be entirely accounted for by the known principles of physics and chemistry.

The result is that modern biology accepted the concept of information but not the concept of meaning, because meaning does not belong to physical theory. But meaning is inseparable from semiosis, so how can we explain the existence of symbols and codes in the cell by relying only on known physical principles?

This is the problem that Pattee set out to resolve, and his starting point was the idea that physical theory does not consist only of physical laws. Influenced by Michael Polanyi, he saw that it consists of laws *plus* initial conditions and boundary conditions that are often referred to as *constraints*. This had been known since Newton's time, but physicists had consistently assumed that laws are fundamental whereas constraints have only an accessory role. Reality, however, turned out to be very different.

Murray Gell-Mann (1994) underlined that the effective complexity of the universe receives only a small contribution from the fundamental laws. The rest comes from 'frozen accidents', which are precisely the result of constraints. All planets, for example, are formed according to universal physical laws, and yet they are all different. Their individual features are due to the particular constraints of their development, and the distinction between laws and constraints is so important that Eugene Wigner (1964) called it "*Newton's greatest discovery*".

In this novel theoretical framework, where laws and constraints have equally fundamental roles, Pattee argued that information and codes are perfectly compatible with normal physical theory because they qualify as constraints. The rules of a code, for example, are limitations that drastically reduce the number of possibilities and can be regarded therefore as natural constraints. In a similar way, Claude Shannon underlined that information is obtained whenever uncertainty is reduced, and concluded that the notions of information and constraint are interchangeable (Shannon 1948).

The solution proposed by Pattee, in short, is that signs and codes do not require new laws of physics, because they are *a special type of constraints* and constraints are an integral part of normal physical theory. The argument is developed in three logical steps: (1) life requires evolvable self-reproduction (a biological principle), (2) evolvability requires symbolic control of self-reproduction (von Neumann), and (3) symbols and codes are special types of constraints (Pattee).

This proposal is undoubtedly a form of biosemiotics, because it states that semiosis exists in every living cell, and since it is based on the idea that signs and codes are physical constraints, it can be referred to as *physical-constraint biosemiotics*, or, more simply, as *physical biosemiotics* (Pattee himself, in a private correspondence with the author, has accepted that this is an adequate name for his approach).

## 9.3 Darwinian Biosemiotics

According to the Modern Synthesis, life does not require new laws of physics, but does require a new biological principle because it is based on natural selection, and this is a process that does not exist in the inanimate world. The principle of natural selection is unique to life, and represents *something extra* in respect to the laws of physics. The Modern Synthesis, in short, takes as explanatory principles not the laws of physics alone, but the laws of physics *plus* the principle of natural selection.

Ernst Mayr, one of the architects of the Modern Synthesis, has repeatedly emphasized that biology has necessarily a wider framework than physics: "... The explanatory equipment of the physical sciences is insufficient to explain the interplay between historically acquired information and the responses of the genetic program to the physical world.... This is why it is just as impossible to include biology in physics as it is to include physics in geometry" (Mayr 1982).

The addition of natural selection to the laws of physics, was accepted by Howard Pattee, and this is a step that amounts to an extended version of his theory. Pattee underlined that there is no contrast between physical biosemiotics and natural selection, and we can therefore put them together. In this way, by adding a mechanism of evolution to physical biosemiotics, we obtain *evolutionary physical biosemiotics*, and since the mechanism in question is natural selection we can rightly call it *Darwinian biosemiotics.*

This view is in agreement with the Modern Synthesis because it accepts its basic concepts, in particular the idea that natural selection is the sole mechanism of evolution. Darwinian biosemiotics, in other words, is physical biosemiotics plus natural selection, and can be regarded as the evolutionary version of physical biosemiotics.

In both cases, the Pattee approach introduces semiosis into the Modern Synthesis and represents, to all effects, an updating of the Synthesis itself, an extension that allows it to incorporate the concept of semiosis and recognize at last the existence of '*symbols that control matter*' in every living cell.

Another approach that can be referred to as Darwinian Biosemiotics is that proposed by Terrence Deacon with three distinct models that attempt a synthesis of the views of Darwin and Peirce. The first is the *Symbolic Species* model on the origin of language (Deacon 1997), the second is the *autocell hypothesis* on the origin of life (Deacon 2006), and the third is a comparison of natural selection with the concepts of entropy proposed by Boltzmann and by Shannon (Deacon 2007, 2008).

## 9.4 The Peirce Framework

Charles Sanders Peirce (1839–1914) was an American polymath, educated in chemistry, who gave lasting contributions to philosophy, logic and semiotics. In philosophy he was a founder of *pragmatism* (or *pragmaticism* as he called it), and in logic he was the first who went beyond Aristotle by adding the new category of *abduction* to the classical Aristotelian categories of *induction* and *deduction* (abduction is an 'extrapolation from limited data', an operation by which a result is reached by 'jumping to conclusions'). But his greatest contribution was in semiotics, the *doctrine of signs*, a field that he virtually rescued from oblivion after centuries of neglect (Favareau 2007). His work in this field touches countless issues, but here we will briefly summarize only three of them: the classification of signs, the definition of semiosis, and the idea of 'extended mind'.

1. *The classification of signs*

   A sign has been traditionally defined as '*something that stands for something else*', and in antiquity a fundamental distinction was made between *cultural signs* and *natural signs*. Peirce underlined that a detailed investigation allows us to classify the signs first into 10 and then into 66 different classes (Sanders 1970; Parker 1998; Short 2007). At the same time, he broke with tradition and claimed that there is no fundamental difference between cultural signs and natural signs on the ground that they all require acts of *interpretation*.

2. *The definition of semiosis*
   Peirce maintained that semiosis is always a triadic process because it necessarily requires an interpretant. "*By 'semiosis' I mean an action, or influence, which involves a cooperation of three subjects, such as a Sign, its Object, and its Interpretant, this triadic influence not being in any way resolvable into actions between pairs*" (Peirce 1906).
3. *The idea of 'extended mind'*
   Animals have feelings and instincts and the ability to interpret the world, and all this tells us that they have 'mind'. But can mind exist without nervous systems? The evidence that animals have mind is not enough to answer this question, but Peirce claimed that it becomes enough when we integrate it with the *principle of continuity*. This principle tells us that everything in evolution comes from less developed forms, so there have necessarily been more primitive forms of mind in all living systems that came before animals, and in their precursors as well. This is why Peirce embraced the view that an *extended* form of mind exists throughout the universe, and concluded that "... *the entire universe – not merely the universe of existents, but the universe which we are all accustomed to refer to as 'the truth' – all this universe is perfused with signs, if it is not composed exclusively of signs.*" (Peirce 1906).

The theoretical framework of Peirce, in conclusion, is build on the idea that semiosis is everywhere in the universe, a view that is known as *pansemiosis* or *physiosemiosis*. It is a modern version of the ancient doctrine of *hylozoism*, the idea that 'matter is pregnant with life', that 'matter is exhausted mind'.

It could be pointed out that the principle of continuity works both ways: it is equally legitimate to say that '*it is mind all the way down*' as to claim that '*it is chemistry all the way up*'. In both cases it is the principle of continuity that leads to a solution, but it does not tell us what this solution actually is.

## 9.5 Zoosemiotics

The idea that animals have feelings, psychologies and even minds has been entertained in various ways throughout the centuries, but for a long time it has been virtually taken for granted that only man is capable of *semiosis*, i.e., that only man makes use of *signs*. This idea was explicitly challenged for the first time in 1963, when Thomas Sebeok suggested that animal communication is also based on signs and proposed the term *zoosemiotics* for the new science of *animal semiosis* (Sebeok 1963, 1972).

That proposal set Sebeok on a long search for evidence of semiosis in various fields of the life sciences, and eventually the hunt paid off. In the last account of his long chase, Sebeok (2001) declared that he got the crucial clue from biologist Jakob von Uexküll (1864–1944) by reading, in 1976, the original German edition of *Theoretische Biologie* (1928).

The great insight of that book is that animals can play, cheat, threaten, court and act, all of which suggests that they are *interpreting* what goes on around them, and not just reacting automatically as pre-programmed puppets. But is the ability to interpret the world enough to prove that animals are capable of semiosis? The answer depends of course on what we mean by semiosis, and on this point there were, in the 1960s, two major schools of thought, one founded by the Swiss linguist Ferdinand de Saussure (1916) and the other by the American philosopher Charles Sanders Peirce (1906).

The main difference between the two schools is that Saussure defined the sign as a dual entity, a combination of *signifier* and *signified*, whereas Peirce insisted that it is a triadic relationship between a *sign vehicle*, an *object* and an *interpretant*. According to Peirce, any act of signification, i.e., any *semiosis*, is necessarily a process of *interpretation*, and this is the view that Sebeok, a pupil of Charles Morris, came to regard as the true definition of semiosis.

The combination of Uexküll's evidence that animals are 'interpreters' with the Peirce's concept that 'interpretation' is the hallmark of semiosis allowed Sebeok to conclude that animals are indeed semiotic creatures. Sebeok, furthermore, was able to account for the differences that exist between human and animal semiosis by pointing to the existence of three different types of signs that Peirce (1906) had called *icons*, *indexes* and *symbols*.

1. Icons are signs that represent objects in virtue of a *similarity* that exists between them, and similarities are useful because they allow animals to recognize objects from a few general features.
2. Indexes are signs that represent objects because of a *physical link* between them. Footprints, for example, are signs of preceding animals, smoke is a sign of fire, and so on. Indexes are the basic tools of learning because they allow animals to infer the existence of something from a few physical traces of something else.
3. Symbols are signs that represent objects in virtue of *arbitrary* links that are established between them. There is no similarity and no physical link between a name and a person, for example, or between a flag and a country. Symbols allow us to make arbitrary associations and build mental images of hypothetical events (projects), of abstracts things (numbers), and even of non-existing things (unicorns).

It is an experimental fact that human culture, and language in particular, is massively dependent on symbols, whereas animal communication is based almost exclusively on icons and indexes. It is true that some examples of symbolic activity have been reported in animals, but in no way can be regarded as primitive languages or intermediate stages toward language. A pervasive and systematic use of symbols is indeed uniquely human, but icons and indexes too are signs and this is what allowed Sebeok to conclude that animal semiosis is a reality.

## 9.6 Peircean Biosemiotics

After the pioneering work of Jacob von Uexküll, the study of animal behaviour has grown into a research field in its own right and has proved with countless examples that animals are indeed interpreters, in species-specific ways, of what goes on in the world around them. In the Peirce framework, on the other hand, interpretation is the defining feature of semioisis, and this allowed Sebeok to conclude that semiosis exists in all animals and not only in man. But why should semiosis be confined to animals? What about other living creatures? By the 1980s, there were at least three lines of evidence that suggested the presence of semiosis beyond the animal kingdom.

1. Marcel Florkin (1974) underlined that the genetic code is proof that semiosis exists at the molecular level and its study represents a new field of research to which he gave the name of *molecular biosemiotics*.
2. Martin Krampen (1981) pointed out that plants too make use of signs and engage in a plant-specific type of communication that he called *phytosemiosis*.
3. Free-living single cells make up the great majority of the living world, and countless studies have shown that their behaviour is *context-dependent*, in the sense that they can react in different ways to different environmental conditions. In this respect they appear to be similar to animals, and that suggested that the context-dependent behaviour of microorganisms is evidence that they have the ability to interpret the signals from the environment.

This is why Sebeok, in 1983, proposed to a group of colleagues to investigate the possibility that Peirce's semiotics could become a paradigm for the whole of biology. The result of that teamwork was published the following year with the title '*A Semiotic Perspective on the Sciences: Steps toward a New Paradigm*' (Anderson et al. 1984) and became the 'manifesto' of a synthesis between biology and semiotics that shortly afterwards was referred to as *biosemiotics*.

A few years later Sebeok organized a series of international meetings at Glottertal, near Freiburg, and it was in one of those meetings, in 1990, that he met Jesper Hoffmeyer, a Danish biochemist who had founded a *Society for the Semiotics of Nature* in Copenhagen. Shortly afterwards, in 1992, came the encounter with Kalevi Kull, who was organizing the *Jakob von Uexküll Center* in Tartu, Estonia.

Hoffmeyer and Kull were biologists, not semioticians, and their joining in turned biosemiotics into a fully interdisciplinary enterprise. It also marked the transition to a younger generation, and perhaps it is fair to say that the passing of the testimonial from Sebeok to Hoffmeyer took place in 2001, when the first Gathering in biosemiotics was organized by Hoffmeyer in Copenhagen (Kull 2001; Favareau 2007).

The formal definition of semiosis in terms of interpretation appeared in the treatise edited by Posner, Robering and Sebeok in 1997, and was expressed in these terms: "*The necessary and sufficient condition for something to be a semiosis is that A interprets B as representing C, where A is the interpretant, B is an object and C is the meaning that A assigns to B*" (Posner et al. 1997, p. 4).

As a result of Sebeok's influence, it has been taken almost for granted, in biosemiotics, that interpretation is the defining feature of both semiosis and life. Sebeok expressed this concept in no uncertain terms: "*Because there can be no semiosis without interpretability – surely life's cardinal propensity – semiosis presupposes the axiomatic identity of the semiosphere with the biosphere*" (Sebeok 2001, p. 68).

The identification of biology with Peircean semiosis was also accepted by Jesper Hoffmeyer, who stated that "*the basic unit of life is the sign, not the molecule*" (Hoffmeyer 1996). A similar concept was expressed by Kalevi Kull: "*Sign science and life science are coextensive ... semiotics is biology and biology is semiotics*" (Kull 2001).

## 9.7 Hermeneutic Biosemiotics

In *Readers of the Book of Life*, Anton Markoš (2002) proposed a view of life that is based on interpretation, like *Peircean biosemiotics*, but was inspired by the hermeneutic philosophy of Martin Heidegger and Hans Georg Gadamer, and has become known as *biohermeneutics* or *hermeneutic biosemiotics*.

The starting point in this *approach* is the problem of 'novelty'. Do genuine novelties exist in nature? Did real novelties appear in the history of life? In classical physics, as formulated, for example, by Laplace, novelty was regarded as a complete illusion, and although this extreme form of determinism has been abandoned by modern science, the idea that nothing really new happens in the world is still with us. It comes from the idea that everything is subject to the immutable laws of nature, and must therefore be the predictable result of such laws. There can be *change* in the course of time, but only relative change, not absolute novelties.

Against this view, Markoš underlined that in human affairs we do observe real change, since our history is ruled by *contingency*, and entities like literature and poetry show that creativity does exist in the world. From this he maintained that a creative view of history can be extended to all living creatures, and argued that this is precisely what Darwin's revolution was about. It was the introduction of contingency in the history of life, the idea that all living organisms, and not just humans, are *subjects*, individual *agents* that act on the world and take care of themselves. Darwin did pay lip service to the determinism of classical physics, but what he was saying is that evolution is but a long sequence of '*just so stories*', not an orderly unfolding of events dictated by immutable laws (Markoš et al. 2009).

According to Markoš, the present version of Darwinism that is known as Modern Synthesis, is a substantial distortion of the original view of Darwin, because it is an attempt to explain the irrationality of history with the rational combination and recombination of chemical entities. Cultural terms like information and meaning have been extended to the whole living world, but have suffered a drastic degradation in the process. Information has become an expression of statistical probabilities, and meaning has been excluded *tout court* from science.

Darwin showed that the history of life is as contingent as the history of man, and Heidegger showed that man can create genuine novelties because he is *interpreting* what goes on in the world. This is enough, according to Markoš, to say that all living creatures are interpreting subjects, and that all novelties in the history of life were brought into being by acts of interpretation.

He concluded that we should go back to the basic approach of the humanities and study the history of life as a *narrative*, precisely as we do with the history of human affairs. Hermeneutic biosemiotics calls in this way for a radical new synthesis of biology and semiotics, a synthesis where biology transcends the objective world of science and becomes an extension of the humanities.

## 9.8 A Book Review for Sebeok

In March 2001 I sent to Thomas Sebeok the first version of '*The Organic Codes*', a manuscript where I pointed out that there are many organic codes in Nature and that their appearance in the history of life was associated with the great events of macroevolution. I also argued that the existence of the organic codes requires that we introduce in biology not only the concept of *information*, but also the concept of *meaning*.

Sebeok kindly acknowledged the manuscript and a few weeks later invited me to review a special issue of *Semiotica* edited by Kalevi Kull and entitled '*Jakob von Uexküll: A paradigm for biology and semiotics*' (Kull 2001). I accepted with enthusiasm, at first, but soon became aware of a sharp contrast between our positions. In my book I had stressed that organic meaning comes from coding, whereas all contributors to the special issue were endorsing the Peircean view that meaning is always produced by a process of interpretation. Such a contrast could hardly be ignored, and in my review I pointed out that the crucial point is the position that we take on the issue of *mechanism*.

The endorsement of a non-mechanistic approach to life was indeed a constant underlying theme of the special issue, and biosemiotics was portrayed as the crowning achievement of the idealistic tradition in biology "... *a lineage involving scholars like Goethe, Saint-Hilaire, von Baer, d'Arcy Thompson, Spemann, Driesch, Waddington, Brian Goodwin, René Thom and Stuart Kauffman*" (Kull 2001, p. 97). I argued instead that the existence of organic codes and organic meaning in nature are scientific problems that can and should be investigated with the classical method of science, i.e., with the mechanistic approach of model building.

More precisely, I underlined that 'understanding' something means explaining it with a model that we are familiar with, and a machine gives us an immediate sense of familiarity. When we see it working before our eyes, we feel that we 'know' it. Actually, we do not even need to build a machine to get this feeling. A description is enough, and so a machine is often a *model*, or even an *algorithm*. A model, furthermore, does not necessarily have a mathematical form. Natural selection, for example, is a mechanistic model which is entirely expressed in words.

Mechanism, in short, is virtually equivalent to the scientific method. The difference is that the *hypotheses* of the scientific method are replaced by *models*, i.e., by descriptions of fully functional working systems. Mechanism, in other words, is *scientific modelling*.

I concluded my review by saying that "*the first two main points of this special issue – the making of biosemiotics and the recovery of Jakob von Uexküll from oblivion – come out with clarity and force, and are definitely a success. Unfortunately, however, there is also a third less happy theme that is developed throughout the volume. The endorsement of non-mechanism, or qualitative organicism, is in my opinion the first serious mistake of the young field of biosemiotics. Indeed it is the one drawback that can prevent it from growing into a true science. I must conclude therefore that biosemiotics has not yet come of age, but I do hope that this criticism is taken for what it is: a diagnosis that is supposed not to hurt but to help*" (Barbieri 2002, pp. 294–295).

## 9.9 A Pluralistic Enterprise

I sent my review to Thomas Sebeok in August 2001, saying that I had not been able to write an impartial report and therefore that I would not be surprised if he turned it down. Surprisingly, however, Sebeok accepted it, and I received a copyright transfer form to fill in. That gave me the idea to test his determination so I answered that I needed to keep the copyright to myself for a forthcoming book. Since he had been taken ill, it was his wife Jean who replied and wrote "*he has made some rare exceptions to the copyright rule when necessary, and he would be willing to do so in this case*". That convinced me that Sebeok *wanted* to publish that review, and this meant that he was not against, *in principle*, to a mechanistic approach to the problem of meaning.

Sebeok died a few months later, on December 21, 2001, and my review appeared in *Semiotica* in the following year (Barbieri 2002). Personally, I took it as an invitation to join the biosemiotic community and to argue in favour of a mechanistic approach from within that community. I decided therefore to give it a try and asked to take part in the second Gathering in Biosemiotics that was going to take place in Tartu, Estonia, in June 2002. There I found a genuine attention to novel ideas and a willingness to discuss them without the constraint of ideological principles. Gone were the triumphal tones and the neo-vitalistic declarations of the Special Issue, and I came away with the feeling that nothing had been settled yet, that everything was on the move. I had the same impression a year later, at the third Gathering that took place in Copenhagen, and realized that a dialogue could start between us over the problem of meaning in nature. At the same time, I became aware that our discussions were not enough. We needed to reach out to a larger audience and that is why, in March 2004, I proposed to create a new journal specifically dedicated to biosemiotics.

The agreement was reached in June 2004, at the fourth Gathering organized by Anton Markoš in Prague. Jesper Hoffmeyer, Claus Emmeche, Kalevi Kull, Anton

Markoš and myself met in a pub and decided that what was uniting us – the introduction of meaning in biology – was far more important than our divisions. Up until then, I had been referring to the study of biological meaning as *semantic biology*, whereas Markoš was calling it *biohermeneuthics*, but we accepted to give up those favourite names of ours and to adopt the term *biosemiotics* that Sebeok had been campaigning for with so much passion and vigour. At the same time, I underlined that our differences did not have to be suppressed and different schools of biosemiotics could and should coexist. The building of biosemiotics, in other words, had to be *a pluralistic enterprise*, and we all agreed on this point.

The practicalities of looking for a publisher, testing the market and collecting a critical mass of papers took a few years, but eventually I signed a contract with Springer as editor-in-chief of *Biosemiotics* and a second contract as co-editor, with Jesper Hoffmeyer, of a Springer Book Series in Biosemiotics. Journal and Book Series started publishing in 2008, and have appeared regularly ever since.

My editorial position, on the other hand, was not preventing me from developing the scientific approach to the problem of meaning, and in particular the idea that organic meaning is produced by coding, not by interpretation. This is the key idea of an approach that started with the name of *Code Biosemiotics*, and later evolved into *Code Biology*, as I shall describe in the rest of this chapter.

## 9.10 Organic Semiosis

A code is a set of rules that establish a correspondence between the objects of two independent worlds. Let us notice now that establishing a correspondence between, say, object 1 and object 2, is equivalent to saying that object 2 is the meaning of object 1. In the Morse code, for example, the rule that 'dot-dash' corresponds to the letter 'A', is equivalent to saying that letter 'A' is the meaning of 'dot-dash'. To an English speaker, the mental object of the sound 'apple' is associated to the mental representation of the fruit 'apple', and this is equivalent to saying that that fruit is the meaning of that sound.

By the same token, the rule of the genetic code that a group of three nucleotides (a codon) corresponds to an amino acid is equivalent to saying that that amino acid is the *organic meaning* of that codon. Anywhere there is a code, be it in the mental or in the organic world, there is meaning (Barbieri 2003). All we need to keep in mind is that *meaning is a mental entity when the code is between mental objects, and an organic entity when the code is between organic molecules*.

This tells us that meaning is often produced by codes, but we know that in animals it is also produced by processes of interpretation, and this suggests that that there are at least two distinct types of semiosis in life, one based on coding and one based on interpretation.

Thomas Sebeok extended semiosis from animals to all living creatures on the ground that free-living single cells like bacteria and protozoa have a *context-*

*dependent* behaviour, and apparently he was convinced that such a behaviour requires the ability to interpret the signals from the environment.

In reality, the context-dependent behaviour of single cells is accounted for much more naturally by the combination of two or more organic codes. A context-dependent behaviour means a context-dependent expression of genes, and this is obtained simply by linking gene expression to signal transduction, i.e., by coupling the genetic code with a signal transduction code. It takes only two context-free codes, in short, to produce a context-dependent behaviour. Coding and decoding, on the other hand, are far simpler than interpretation, and there is no need to assume anything more complicated than that in free-living single cells, especially in those that appeared at the beginning of the history of life.

For all their outstanding abilities in coding and decoding, however, single cells do not build internal representations of the world and therefore cannot interpret them. They are sensitive to light, but do not 'see'; they react to sounds but do not 'hear'; they detect hormones but do not 'smell' and do not 'taste' them. It takes the cooperation of many cells which have undertaken specific processes of differentiation to allow a living system to see, hear, smell and taste, so it is only multicellular creatures that have these experiences.

Free-living single cells are semiotic creatures because they make use of signs, but their semiosis is based exclusively on organic codes, and for this reason it has been referred to as *organic semiosis*. There simply is no evidence of interpretation at the molecular level. One of the best proofs of this comes from the fact that the genetic code has been highly conserved in all living organisms and in all environments ever since the origin of life, which clearly means that it does not depend on interpretation.

## 9.11 Animal Semiosis

The origin of animals is described as the origin of multicellular creatures that evolved a nervous system and acquired what we call feelings, sensations, perceptions, mental images and so on. The transformation of the signals received by the sense organs into mental images, or high-level neural states, is based on sets of rules that are often referred to as *neural codes*, because there are no necessary connections between sensory inputs and mental, or neural, images. This suggests that the appearance of different neural codes in the evolution of the brain is comparable to the appearance of different organic codes in the evolution of the cell, both starting from a virtually universal original code.

Such a scenario, however, turns out to be only partially true. The increase in the number of neural codes would explain the increase in complexity that took place in brain evolution, but would produce animals whose behaviour is almost entirely pre-programmed, or *hardwired*, because codes are based on fixed rules. As a matter of fact, many animals (for example fishes) do have virtually hardwired reactions, and in those cases animal behaviour is indeed largely accounted for by neural codes only. At the same time, however, it is a fact that some animals evolved the ability to *interpret* what goes on in the world around them, and this skill is a true evolutionary novelty, something that is not reducible to coding alone.

Interpretation is the ability to reach a conclusion from sensory inputs whose result can vary according to circumstances, memory, experience and learning. In a way, it is the ability to 'jump-to-conclusions', so to speak, from a limited number of data, with results that may not be perfect but are good enough for survival purposes. This 'extrapolation from limited data' is not reducible to the classical Aristotelian categories of induction and deduction, and for this reason Charles Peirce called it *abduction*. It is a new logical category, and the ability to interpret the world appears to be based precisely on that logic. Interpretation, in short, is a form of semiosis because it involves signs and meanings, but it is different from organic semiosis because it is based not only on coding but also on abduction.

Interpretation, furthermore, is a faculty that did not appeared full blown in animals. It evolved in stages, and we can still see intermediate forms of its historical evolution. Some animals (like snakes), for example, stop chasing a prey when it disappears from sight, whereas others (like mammals) deduce that the prey has temporarily been hidden by an obstacle and continue chasing it. Some can even learn to follow the footsteps of a prey, which reveals a still higher degree of abstraction (Barbieri 2011).

What animals interpret, furthermore, is not the world but *representations* of the world, and neural representations are formed by neural networks made of many different types of cells, which means that interpretation can exist only in multicellular systems.

The evolution from single cells to animals, in conclusion, was far more than an increase in growth and complexity. It was a true macroevolution that gave origin to absolute novelties, to entities that did not exist before such as perceptions, feelings and the ability to interpret what goes on in the surrounding world. That is what divided animals from single cells, and that is why there are at least two distinct types of semiosis in living systems: one that is based on coding and decoding (*organic semiosis*) and one that is based on interpretation (*animal semiosis*).

## 9.12 Sense and Reference

The existence of two distinct types of semiosis, one based on coding and the other on interpretation, does have a logical basis in two distinct concepts of meaning that were first introduced by Gottlob Frege in 1892. The meaning of a sign is, by definition, what the sign stands for, but this can be of two very different types. A sign can stand for an object that is either 'external' or 'internal' to the system. As an example, Frege pointed out that the terms *Morning Star* and *Evening Star* both indicate the planet Venus, and have therefore the same 'external' meaning, but very different 'internal' meanings. The Evening Star is visible at dusk and was related to end-of-the-day events, whereas the Morning Star was related to the rise of the sun and to the beginning of a new day (Frege 1892).

In order to differentiate between internal and external meanings, Frege gave them two different names. He called them respectively '*sense*' and '*reference*'. Other authors have adopted a different terminology, and have used terms like 'intension' and 'extension', or 'connotation' and 'denotation'. The point, however, is the same. All entities associated to signs qualify as meanings, but some may refer to internal entities (sense) and others to outside objects (reference).

After Frege, it has been taken virtually for granted that all signs have both types of meanings, but this is not always the case. Proteins, for example, are built according to the genetic code, and this is a set of exclusively internal rules. Proteins do not have other meanings apart from those that come from genes, and this implies that the apparatus of protein synthesis is a system of signs which have sense but not reference, i.e., signs which have internal meanings but not external ones. The same is true for all the other organic codes that exist in single cells.

The human brain, on the other hand, receives signals from the environment and uses them to produce internal models of the world, models that are the *reference*, the external meanings, that the brain associates to the incoming signals. The brain, however, is also giving internal meanings to signs, and must integrate them with the external meanings in order to allow the body to react to the incoming signals. This makes us understand why the brain (or the mind) must necessarily use *two* types of meanings. If the mind could act *directly* on the environmental signals, there would be no need of external meanings, but that is not what happens. The mind can only act on *representations* of the world and that is why it must use signs which have both internal and external meanings, i.e., sense and reference.

There is, in short, a real difference between the semiosis of single cells and that of animals. A semiosis based on coding and decoding requires only one type of meaning (sense) whereas a semiosis based on interpretation requires two distinct types of meaning (sense and reference).

## 9.13 The Worlds of Semiosis

According to the classical doctrine of semiosis developed by Aristotle, Augustine and Aquinas, signs are divided into two great categories – natural signs and cultural signs – and this for two different reasons. One is because they derive either from nature (*signa ex natura*) or from culture (*signa ex cultura*). The other is because they are either symptoms (*signa symptomatica*) or symbols (*signa symbolica*). If we explicitly mention both these characteristics, signs are defined in the following way:

1. the natural signs are *signa symptomatica ex natura*, and
2. the cultural signs are *signa symbolica ex cultura*.

   This distinction means that there are two types of semiosis in the world: *natural semiosis* is what we observe in animals whereas *cultural semiosis* is an exclusively human phenomenon.

The discovery of the genetic code came as a bolt from the blue precisely because it revealed the existence of a *third* category of signs, a category that all great thinkers of the past had not been able to predict: the existence of *symbols that come from nature, not from culture*. In addition to the two classical categories, therefore, we now have a third one:

3. the organic signs are *signa symbolica ex natura*.

This is the immense novelty of the genetic code: it brought to light an entirely new type of semiosis. This breakthrough was later generalized by the discovery of many other organic codes, and the result is that today we have three distinct types of semiosis in Nature.

(a) Standard scientific tests prove that organic codes exist in single cells and this tells us that there is in Nature a form of semiosis that is based on coding and decoding (*organic semiosis*).
(b) There is good evidence that some ancestral animals started evolving the ability to interpret what goes on in the world around them, and this means that there is in Nature a second form of semiosis that is based on coding and interpretation (*animal semiosis*).
(c) Finally, we have impressive evidence that a third type of semiosis evolved in our species, a semiosis that is based on coding, interpretation and language (*human semiosis*).

The three types are often referred to as *biosemiosis, zoosemiosis* and *anthroposemiosis*, but it must be underlined that there is no universal consensus about the differences that exist between them. In the Peirce framework, in particular, it is accepted that semiosis can take on different forms, but it is claimed that there is no fundamental discontinuity between them because they all require interpretation.

This, however, is precisely the point that we cannot accept *a priori*. We do know that animals are capable of interpretation, but this is not enough to say that the same is true in all living systems, and in particular in single cells. The key issue, in other words, is the *nature* of the differences between the three known forms of semiosis. Are they fundamentally continuous processes or do *real discontinuities* exist between them?

## 9.14 Two Types of Biosemiotics

The *manifesto* written by Thomas Sebeok and colleagues (Anderson et al. 1984) declared that the semiotics of Peirce is a "*paradigm for biology*", a theoretical framework that illuminates the life sciences because it provides them with entirely new concepts. Hence the project of introducing Peircean concepts in biology, a project that a few years later Sebeok referred to as *biosemiotics*, the synthesis of biology and semiotics (Sebeok and Umiker-Sebeok 1992).

Ever since 2001, when Sebeok asked me to review a special issue of *Semiotica* dedicated to Jacob von Uexküll, I have voiced a specific criticism to that project. I repeatedly argued that a synthesis of biology and semiotics can and should be a *scientific* research project where the Peirce model can be *tested* rather than being taken for granted. I underlined, in other words, that there are two types of biosemiotics before us: one is the extension of the Peirce's model to all living crealtures, the other is a scientific approach that aims at *discovering* which semiotic processes actually take place in living systems.

The journal *Biosemiotics* started publishing in 2008, and since then I made annual surveys of the opinions expressed by the authors of papers, by the members of the editorial board and by the participants in the Gatherings in Biosemiotics. The results turned out to be a complete surprise to me. Most biosemioticians acknowledged that what we find in single cells is only coding and decoding, but maintained that cells are nevertheless capable of interpretation *because we can define decoding as a form of interpretation*. There are countless examples of this trick in the literature, and these are just two of them.

(a) In the paper "*What Does It Take to Produce Interpretation?*", Søren Brier and Cliff Joslyn (2013) proposed to solve the problem in this way: "... we can identify *interpretation* in general as any process which encounters a sign and takes it for its meaning in virtue of some code... Thus *a ribosome is an interpreter*. And the right amino acid is its interpretation of some codon."
(b) In the paper "*Anticipatory Functions, Digital-Analog forms and Biosemiotics*", Arnellos et al. (2012) offered a Peircean description of signal transduction and wrote that "... receptors act *as interpreting systems* by coupling to transducers catalytic molecules that trigger the production of another sign inside the cell in response to the extracellular sign."

Here we are then. One can say that cells are interpretive systems simply by adopting an *ad hoc* definition of interpretation. Words, after all, are tools that we employ for our own purposes and we are used to give them multiple meanings, so where is the problem?

The problem is not the use of words, that are indeed tools, but the result that we obtain with them, and in our case, the result is two very different types of biosemiotics.

1. Peircean biosemiotics is essentially a reformulation of known biological processes in Peircean terms. It describes living systems with new words but does not discover anything new about them.
2. Only a scientific approach to biosemiotics can lead to genuine discoveries, but requires that we learn from experiments, not from *ad hoc* definitions, what are the semiotic properties of Nature.

I made regular surveys on these two positions for 4 years, and in the end it turned out that roughly 90 % of the biosemioticians were supporting the Peircean approach. In principle, one could still argue that there are two types of biosemiotics, but in practice it became clear that that was no longer the case.

## 9.15 A New Beginning

In 1962, Thomas Kuhn argued that "… a paradigm corresponds not to a subject matter but rather to a group of *practitioners*", and a few years later a similar thesis was expressed by Mihaly Csikszentmihalyi, who stressed that "a research field includes all the individuals who act as *gatekeepers* to the domain. It is their job to decide whether a new idea or product should be included in the domain" (Csikszentmihalyi 1996).

According to these views, a discipline is what its gatekeepers say it is, and in that sense the identification of biosemiotics with *Peircean biosemiotics* was a perfectly legitimate operation because the board of the *International Society for Biosemiotic Studies* (ISBS) consisted almost exclusively of Peirce followers.

There was also another reason for that conclusion. Semiotics has become today a highly influential discipline and the *Peirce industry* has grown into an impressive enterprise that, like the *Darwin industry* in biology, is producing a steady flow of books, journals, congresses, grants and University positions. Like many other disciplines, on the other hand, semiotics is subdivided into *niches*, one of which is precisely its application to biology. In Academia, as in Nature, all niches tend to be occupied, and this is why the relationship between semiosis and biology has become an increasingly active area in recent years. More than that. Many semioticians openly underlined "how exciting" it was for them the thought of "contributing to the study of life", of "building a new paradigm for biology", and this goes a long way to explain why virtually all semioticians were endorsing the Peircean approach in biosemiotics.

The surveys also brought to light another explanation. The scientists that were supporting the Peircean approach were often using the arguments employed by the supporters of *Intelligent Design*. In retrospect this is hardly surprising, because 'interpretation' is indeed a form of 'intelligence', and Peirce himself promoted the idea that there is an *extended mind* in the Universe. The difference between the two cases is that in Intelligent Design the 'interpreting agency' is outside Nature whereas in Peircean biosemiotics is inside it. The common factor is that in both cases all facts of science are reinterpreted in a *postmodern* framework simply by changing the meaning of words.

By 2012, the surveys made it abundantly clear that most scholars and researchers identify biosemiotics with Peircean biosemiotics, a paradigm that is extending a cultural model of semiosis to the whole of Nature rather than discovering from Nature what biological semiosis actually is. It also became painfully clear that a scientific approach to the semiosis of Nature could not prosper within that framework and its future was seriously at risk.

That is why, at the end of 2012, I resigned as editor-in-chief of *Biosemiotics* and together with 11 colleagues (Jan-Hendrik Hofmeyr, Peter Wills, Almo Farina, Stefan Artmann, Joachim De Beule, Peter Dittrich, Dennis Görlich, Stefan Kühn, Chris Ottolenghi, Liz Swan and Morten Tønnessen) founded the *International Society of Code Biology* (ISCB). We also decided to leave no doubt about the scientific nature of that project, and to this end we explicitly wrote in the constitution of the new society that Code Biology is "the study of all codes of life *with the standard methods of science*".

In 2012, in conclusion, biosemiotics went back to its two historical origins. The majority reaffirmed that biosemiotics is the doctrine that extends the Peirce model of semiosis to all living systems, and a minority group founded the new field of Code Biology in order to keep alive the scientific approach to the problem of meaning that started with the discovery of the genetic code.

# Chapter 10
# Code Biology

Today there are two major paradigms in biology. One is the Modern Synthesis, the framework that has unified the ideas of Darwin and Mendel and has put them on the firm mathematical basis provided by the equations of population genetics. The Modern Synthesis was proposed in the 1930s by Ronald Fischer (1930), Sewall Wright (1931), J.B.S. Haldane (1932), Theodosius Dobzhansky (1937) and others, and then extended, in the 1940s, by Julian Huxley (1942), Ernst Mayr (1942), George Gaylord Simpson (1944), Bernhard Rensch (1947) and others. New extensions have recently been advocated by various authors, in particular by George Williams (1966), Stephen Jay Gould (2002), Gerd Müller (2007), Sean Carroll (2008) and Massimo Pigliucci (2009).

The second theoretical framework is Systems Biology, a paradigm that studies living organisms as self-fabricating systems, in line with the ideas developed by Norbert Wiener (1948), Robert Rosen (1958, 1991), John von Neumann (1966), Mihailo Mesarovic (1968), Ludwig von Bertalanffy (1969), and Humberto Maturana and Francisco Varela (1980).

The Modern Synthesis and Systems Biology are the recent versions of two biological paradigms that have been debated since the early nineteenth century. Traditionally they are known as *functionalism* and *structuralism*, but recently they have also been referred to as *externalism* and *internalism* (Linde Medina 2010). According to functionalism (or externalism) living matter is a fundamentally passive entity that owes its organization to external forces (functions that shape organs) and to an external agent (natural selection). Structuralism (or internalism) is the view that living matter is an intrinsically active entity that is capable of organizing itself from within, with internal processes that are based on mathematical principles and physical laws.

The basic mechanism of the Modern Synthesis is *natural selection*, whereas the central process of Systems Biology is *autopoiesis*, and there is ample evidence that both natural selection and autopoiesis are fundamental biological realities. Molecular biology, however, has uncovered also a third great class of processes at the heart of life. The existence of the genetic code and of many other organic

© Springer International Publishing Switzerland 2015

M. Barbieri, *Code Biology*, DOI 10.1007/978-3-319-14535-8_10

codes in Nature tells us that *coding* is also a biological reality and we need therefore a framework that accounts for it. This framework is *Code biology*, the study of all codes of life, from the genetic code to the codes of culture.

## 10.1 What Is the Cell?

The very existence of life is due to the fact that processes like the copying of genes, the coding of proteins, and the transduction of signals from the environment are executed with extraordinary precision and efficiency. The cell, in other words, performs its operations in an extremely reliable way, and it is legitimate to express this concept by saying that it is a *biological machine*. But what kind of a machine?

Today, it is widely popular to describe the cell as a biological computer, more precisely as a computer-guided *robot*, because it is made of *biological software* and *biological hardware*. In reality, there is a very substantial difference between the two systems. What really divides them is that the codes of a cell come *from within* the system, from an *internal* codemaker that is an integral part of the cell, whereas the codes of a computer come *from without*, from a human codemaker that is *external* to the machine and is made of a totally different stuff. What goes on in a computer, in other words, is human semiosis reduced to automatic operations that can be implemented by a machine. This is why the cell is a semiotic system – a system capable of creating its own codes – whereas a computer, or a robot, is not.

There are, furthermore, other important differences between the organic codes of the cells and the man-made codes of culture. The genetic code, for example, is often compared to the Morse code because both of them consist in mapping rules, but if we take a closer look we discover that there are substantial diversities between them.

(a) One is the fact that the Morse code is perfectly *symmetrical*, or *reversible*. It transforms the letters of the alphabet into dots and dashes and, vice versa, dots-and-dashes into letters of the alphabet. Nothing of the kind takes place in the cell. The genetic code is absolutely *irreversible*. It is a one-way process that goes from genes to proteins and *never* vice versa. The reverse transformation is not just avoided, it is physically impossible.
(b) Another difference is that the messages written in Morse are perfectly equivalent to those written in alphabetic characters. They carry exactly the same information and are simply a different way of expressing the same content. The Morse code, in short, transforms a world of entities into a world of *equivalent* entities. In the case of the genetic code, instead, the situation is totally different. Genes and proteins are not at all equivalent objects, they belong to two completely different worlds.

The cell, in conclusion, can be regarded as a biological machine, but it is very different from our cultural machines, and we need therefore to find out its *logical structure* if we want to understand it.

## 10.2 The von Neumann Machine

In the 1950s, John von Neumann, one of the pioneers of computer science, designed a virtual machine that is capable, in principle, of reproducing itself (von Neumann 1951, 1956, 1966). It is a two-dimensional automaton made of square elements, and von Neumann was able to prove that it can reproduce itself when it contains three fundamental components.

The first is a *Universal Constructor*, a machine that is able to construct any configuration when it is given a description of it. In von Neumann terminology, if A is the universal constructor and φ(A) is its description, A would construct a copy of itself, A′, when it is given its own description. We can express this with the notation:

$$\varphi\,(\mathrm{A}) \rightarrow \mathrm{A} = \mathrm{A}'$$

This is not yet self-reproduction because the description φ(A) has not been replicated. To this purpose it is necessary to add a second component, a *Universal Copier*, B, and its description φ(B). We can now write:

$$\varphi\,(\mathrm{A}+\mathrm{B}) \rightarrow (\mathrm{A}+\mathrm{B}) = \varphi\left(\mathrm{A}'+\mathrm{B}'\right) \rightarrow \left(\mathrm{A}'+\mathrm{B}'\right)$$

At this point, von Neumann realized that each symbol in the description φ(A + B) has *two roles* to play. The first role is to *instruct* A to assemble the new automaton (A′ + B′). The second role is to serve *as a template* so that *B* can create a second copy of the description φ(A + B). Logically it could not perform both of these roles at the same time. Von Neumann introduced therefore a third component, C, an automaton that he called *Controller*, which ensures that the description φ(A + B) is used in the correct role at the correct time, i.e., either to *construct* or to *copy*. But adding the third automaton C von Neumann obtained the scheme of a machine (A + B + C) that is in principle capable of self-reproduction:

$$\varphi\,(\mathrm{A}+\mathrm{B}+\mathrm{C}) \rightarrow (\mathrm{A}+\mathrm{B}+\mathrm{C}) = \varphi\left(\mathrm{A}'+\mathrm{B}'+\mathrm{C}'\right) \rightarrow \left(\mathrm{A}'+\mathrm{B}'+\mathrm{C}'\right)$$

The crucial point in this model is that any additional description of some new component X, when added to the basic description φ will be constructed and incorporated into all future generations:

$$\begin{aligned}\varphi\,(\mathrm{A}+\mathrm{B}+\mathrm{C}+\mathrm{X}) &\rightarrow (\mathrm{A}+\mathrm{B}+\mathrm{C}) = \varphi\left(\mathrm{A}'+\mathrm{B}'+\mathrm{C}'+\mathrm{X}'\right)\\ &\rightarrow \left(\mathrm{A}'+\mathrm{B}'+\mathrm{C}'+\mathrm{X}'\right)\end{aligned}$$

The von Neumann automaton, in other words, is *capable of evolution* because the description of the machine (the software) is separated from the working machine (the hardware) and can therefore change independently from it. The model, furthermore, was describing logical components that later on turned out to be effectively present in the cell. The whole automaton is the phenotype, the description $\varphi$ is the genotype, the universal constructor A is the apparatus of protein synthesis (in particular the ribosomes), the copier B is the transcription enzymes (the polymerases) and the controller C is the regulation enzymes.

It must be underlined, however, that the von Neumann machine is a purely logical construct that takes in no account the *physical nature* of its components. Von Neumann himself was fully aware that logic alone is not enough to explain a physical system, and wrote an explicit warning to this effect:

> By axiomatizing automata in this manner one has thrown half the problem out the window and it may be the more important half. One does not ask the most intriguing, exciting and important questions of why the molecules or aggregates that in nature really occur.... are the sorts of thing they are, why they are essentially very large molecules in some cases, but large aggregations in other cases. (von Neumann 1966, p. 77)

The logical structure of the von Neumann machine, in short, is independent from its physical realization so it is not a complete description of a real system. Something is undoubtedly missing, but what is it?

## 10.3 A Machine Capable of Self-Assembly

The von Neumann machine is a logical architecture that does not depend on the physical characteristics of its components, and this gives us the problem of finding out what is the actual role played in biological systems by their physical 'embodiment'.

In the cell, the crucial point is that genes carry *linear* information whereas the function of proteins is determined by their *three-dimensional* structure. Clearly genes are not carrying all the information that is going to appear in proteins, so where does the missing information come from? Nature's solution to this problem is both simple and extraordinary. The linear information of a gene is used to assemble a linear sequence of amino acids, and then this polypeptide chain undergoes a folding process and assumes a specific three-dimensional form. It is as if one wrote the word 'apple' and then observed the word folding itself up and becoming a real apple.

The informational difference that exists between a linear polypeptide and a three-dimensional protein can be illustrated with a simple example. The linear order of 100 amino acids is specified by 100 coordinates, whereas their three-dimensional arrangement in space requires 300 coordinates (three for each amino acid). Protein folding amounts therefore to adding the 200 missing coordinates to the 100 provided by the genes. And since the complexity of a system is determined by the number of parameters that describe it, it is clear that protein folding is a phenomenon that produces *an increase of complexity*.

The folding of three-dimensional proteins, furthermore, is only the first of a whole series of assemblies. Once formed, many proteins are arranged into supramolecular structures, and these in turn give origin to cell organelles and cell compartments. The enormous amount of information that is stored in the three-dimensional structure of a cell comes from *a nested chain of self-assemblies*, and in all these processes new properties emerge in stages, very much like new structures appear step-by-step in embryonic development. In a von Neumann machine, instead, the programme must provide a complete description of the system and the information of the genotype is identical to the information of the phenotype, which is totally different from what takes place in the cell, where most three-dimensional structures arise by self-assembly.

The von Neumann machine, on the other hand, does account for self-reproduction, and more in general for self-fabrication, or *autopoiesis*, so it is tempting to say that the addition of self-assembly would give us a good model of the cell. It would allows us to say that *the cell has the logical structure of a von Neumann machine that is capable of self-assembly*. But is this enough? Can we really say that autopoiesis and self-assembly is all that we need to have a realistic model of the living cell?

## 10.4 The Concept of Codepoiesis

Modern biology describes the cell as an *autopoietic* system, a *system that fabricates itself* (Hofmeyr 2007). The concept of autopoiesis, or self-fabrication, describes the ability of living systems to produce their own components and eventually to produce copies of themselves. At a closer look, however, we realize that this is not always what takes place in life. In embryonic development, for example, the cells are often *different* from their progenitors and some of them are capable of something that is the very opposite of autopoiesis: they can *suicide* themselves.

It may be pointed out that cells capable of embryonic development appeared very late in evolution and derived from previous autopoietic systems, so it is autopoiesis that we must start with because it was at the beginning of everything. But this is precisely what we *cannot* say. Before the origin of the genetic code, specific proteins could not exist, and the ancestral systems were producing descendants that were inevitably *different* from themselves. They were not autopoietic systems. Autopoiesis, in other words, did not exist before the first cells, so it was not the mechanism that gave origin to them.

Before the origin of the genetic code, the ribonucleoprotein system of the common ancestor was engaged in the process of evolving coding rules and was therefore a *code exploring system*. After the origin of the genetic code, however, the situation changed dramatically. No other modification in coding rules was accepted and the system in question became a *code conservation system*. Another part of the system, however, maintained the potential to evolve other coding rules and behaved as a new *code exploring*, or *code generating, system*. In the early *Eukarya*, for example, the cells had *a code conservation part* for the genetic code, but also a *code exploring part* for the splicing code.

This tells us something important about life. The origin of the first cells was based on the ability of the ancestral systems to *generate* the rules of the genetic code, and the subsequent evolution of the cells was based on two complementary processes: one was the *generation* of new organic codes and the other was the *conservation* of the existing ones. Taken together, these two processes are the two sides of a biological phenomenon that has been referred to as *codepoiesis* (Barbieri 2012).

The ancestral systems that gave origin to the first cells were not autopoietic systems but they had to be codepoietic systems. And all cells that came after them were not always engaged in autopoiesis but were inevitably engaged in codepoiesis. What is always and necessarily present in all living systems, in other words, is codepoiesis, and this gives us a new model of the cell that can be expressed in this way: '*the cell is a codepoietic system, i.e., a system that is capable of creating and conserving its own codes*'.

The von Neumann machine can, in principle, account for autopoiesis because it contains a Universal Constructor, but does not account for codepoiesis, and this is why it is not a realistic model of the cell. We realize in this way that the deep logic of life – the logical structure of the cell – is codepoiesis, not autopoiesis.

## 10.5 Extensions of the Modern Synthesis

The Modern Synthesis of the 1930s and 1940s has unified the ideas of Darwin and Mendel and has put them on the firm mathematical basis provided by the equations of population genetics. These equations have shown that Mendelian inheritance is a conservative mechanism that tends to keep constant the gene frequencies in large populations. Evolution, however, is about change and there are various processes that can change gene frequencies in the course of time. Mutation, migration, random drift and selection can all lead to long term modifications but only one of them – natural selection – allows the organisms to adapt to the environment. Natural selection, in short, is the only mechanism that leads to *adaptive* evolution, and to many biologists this is all that matters. Implicit in this concept is the idea that natural selection operates only *indirectly* on genes because its direct target is the organism since it is whole bodies that compete, mate, leave behind more or less descendants and in the long run adapt to the environment.

In 1964, however, William Hamilton published a study on the genetic basis of behaviour which showed that natural selection can act *directly* on genes. A gene for altruistic behaviour, for example, can induce an individual to sacrifice himself for the community and yet that gene can be favoured by natural selection if it allows the survival of a significant number of relatives that are carrying it (Hamilton 1964). In this case, natural selection is sacrificing the organism in order to propagate the gene among its relatives (*kin selection*), which suggests that the real target of such selection is the gene not the organism.

In 1966, George Williams showed that this conclusion can be extended from the genes of behaviour to *all* genes (Williams 1966) and formulated what has become known as the *gene-centred* view of evolution, a view that 10 years later was successfully popularized by Richard Dawkins in *The Selfish Gene* (Dawkins 1976). This new version of the Modern Synthesis has become highly popular. In *The Major Transitions in Evolution*, for example, John Maynard Smith and Eörs Szathmáry (1995) adopted it *a priori* by declaring, at the very beginning of the book: "we are *committed* to the gene-centred view outlined by Williams (1966) and made still more explicit by Dawkins (1976)".

It must be underlined that this view has *not* been universally accepted in the Darwinian camp. Ernst Mayr and Sewall Wright, for example, did not subscribe to it and remained faithful to Darwin's idea that natural selection acts on individual organisms. Stephen Jay Gould too expressed disagreement, and claimed that natural selection operates at many different levels simultaneously: on genes, cells, organisms, species and higher taxa. This is the *hierarchical view of evolution*, a theory that Gould (2002) regarded as an extension, or a *reformulation*, of the Modern Synthesis that has become necessary because *microevolution* (the origin of species) does not automatically account for *macroevolution* (the origin of higher taxa).

Finally, another extension of the Modern Synthesis has been prompted by the discovery that, against all predictions of population genetics, homologous genes do exist in vastly distant phyla and are responsible for key steps in embryonic development. This latest extension has largely been inspired by the new research field of *Evolutionary Developmental Biology* (*EvoDevo*), and has been advocated in particular by Gerd Müller (2007), Sean Carroll (2008), Massimo Pigliucci (2009) and Pigliucci and Müller (2010).

In the last few decades, in conclusion, there have been at least three major extensions of the Modern Synthesis, which clearly shows that the field is in an active state of development. This, however, does not mean that an extended Modern Synthesis is enough to solve the problem of the mechanisms of evolution.

## 10.6 Mechanisms of Evolution

The mechanisms of evolution have been one of the most controversial issues in biology and the great debate about them culminated, in the 1930s and 1940s, in the Modern Synthesis, the theoretical framework where natural selection is regarded as virtually the sole mechanism of evolutionary change. But where does natural selection come from?

Let us start from the fact that the copying of the genes is the elementary act that leads to *heredity*. When the process of copying is repeated indefinitely, however, another phenomenon comes into being. Copying mistakes become inevitable and in a world of limited resources not all changes can be implemented, which means that a process of selection is bound to take place. Molecular copying, in short, leads to heredity, and the indefinite repetition of molecular copying in a world of limited resources leads to *natural selection*.

That is how natural selection came into existence. Molecular copying started it and molecular copying has perpetuated it ever since. Which means that *natural selection would be the sole mechanism of evolution if molecular copying were the sole basic mechanism of life.*

The discovery of the genetic code, however, has proved that there are *two* distinct molecular mechanisms at the basis of life, the *copying* of the genes and the *coding* of proteins. The discovery of other organic codes, furthermore, allows us to generalize this conclusion because it proves that coding is not limited to protein synthesis. Life, in other words, is not based on copying alone. It is based on copying and coding, and these two molecular mechanisms give origin to two distinct mechanisms of evolution because an evolutionary mechanism is but the long term result of a fundamental molecular mechanism.

More precisely, the existence of copying and coding at the molecular level means that there are two distinct types of evolutionary change: *evolution by natural selection*, based on copying, and *evolution by natural conventions*, based on coding (Barbieri 1985, 2003).

But can we prove it? The only way to do it is to prove that the two mechanisms are fundamentally different, i.e., that *coding cannot be reduced to copying*. This is therefore our challenge. We can prove that natural selection and natural conventions are two distinct mechanisms of evolution if we prove that copying and coding are two fundamentally different mechanisms of molecular change.

## 10.7 Copying and Coding

Copying and coding are both capable of bringing novelties into the world, but they do it in very different ways. By its very nature, the copying mechanism produces either exact copies or slightly different versions of the copied molecules. This means that natural selection produces new objects by modifying previous ones, i.e., by making objects that are only relatively different from their predecessors. Natural selection, in short, creates *relative* novelties, not absolute ones.

In the case of coding the situation is totally different. The rules of a code are not dictated by physical necessity, and this means that a new code can establish relationships that have never existed before in the Universe. The objects that are assembled by the rules of a new code can have no relationship whatsoever to previous objects. Natural conventions, in short, create *absolute* novelties, not relative ones.

A second difference between the two mechanisms is that copying operates on *individual* molecules, whereas coding involves a *collective* set. The difference between natural selection and natural conventions, in other words, is the difference that exists between individual change and collective change. An example of this difference can be seen in language, whose evolution is due to variations that take place not only at the level of the individual words but also at the level of the collective rules of grammar.

A third difference between copying and coding is that they involve two different entities. A variation in the copying of a gene changes the linear sequence, i.e., the *information* of that gene. A variation in a coding rule, instead, changes the *meaning* of that rule. The great difference that exists between copying and coding, and therefore between natural selection and natural conventions, comes from the difference that exists between 'information' and 'meaning'.

There are, in conclusion, three major differences between copying and coding: (1) copying modifies existing objects whereas coding brings new objects into existence, (2) copying acts on individual objects whereas coding acts on collective sets, and (3) copying is about biological information whereas coding is about biological meaning. Copying and coding, in conclusion, are fundamentally different mechanisms of molecular change, and this makes us realize that evolution was not produced by natural selection alone but *by natural selection and by natural conventions*.

## 10.8 Codes as Constraints

The existence of organic codes in Nature is an experimental fact, but do they fit in the existing theoretical framework? In order to address this problem, we must keep in mind that the great biological theories – on evolution, heredity, embryonic development, etc. – have all been built *before* the discovery of the genetic code and this gives us a problem: how did people account for the genetic code in a theoretical framework that did not contemplate the existence of codes?

The explanation that was given at the time, and that is still widely accepted today, is that the genetic code is a set of *constraints*. This idea was suggested by the distinction that exists in physics between the *laws of nature* and the boundary conditions that are collectively known as 'constraints'. All planets, for example, are formed according to universal physical laws, and yet they are all different because their individual features were determined by the particular local constraints that affected their historical development.

The conclusion that the genetic code is a set of constraints is formally correct because a code is indeed a set of limitations since it consists in a small number of rules that are selected from a virtually unlimited number of possibilities. On top of that, the idea was attractive because physical constraints cannot be changed and this appeared to explain the fact that the genetic code has been 'frozen' since the origin of life.

There is no doubt that physical constraints impose limitations on living systems and it is true that coding rules are constraints, but there are two other arguments that must be taken into account.

1. The first is that the rules of the genetic code are *biological* constraints, not *physical* ones. They are *biologically generated* rules and in no way can be assimilated to physical constraints. This is because the genes of the genetic code are constantly subject, like all other genes, to mutation and neutral drift. They

are in a continuous state of flux and the fact that they have been highly conserved in evolution means that there is a biological mechanism that *actively and continuously* restores their original structure. The conservation of the genetic code, in other words, is not the passive result of physical constraints that somehow 'freeze' it. It can only be the result of an active biological mechanism that is continuously at work, a mechanism that has been referred to as *codepoiesis*.
2. The second argument is that the genetic code is *much more* than a set of constraints. It is the novelty that gave origin to the protein world, to the first macroevolution in the history of life. This means that coding is not only a set of constraints. It is a *generative* mechanism, a mechanism of evolution.

Natural selection is the long term result of the copying of the genes and according to the Modern Synthesis is the sole mechanism of evolution. The discovery of the genetic code proved instead that there are *two* distinct molecular mechanisms at the basis of life, the *copying* of the genes and the *coding* of proteins. This was its real revolutionary message, but it went unnoticed. Biologists did not conclude that copying and coding are two fundamental mechanisms. They concluded instead that copying is a fundamental mechanism and coding is a set of constraints. This is what prevented them from realizing that copying and coding are equally important mechanisms that have *complementary* roles in life.

In a way, this complementarity recalls the relationship that exists between particles and waves, in the sense that one cannot exist without the other and cannot be reduced to the other. But nobody in physics has ever suggested that the existence of particles and waves can be explained by saying that waves are constraints on particles or particles constraints on waves. Particles and waves are complementary realities, and we must accept this as a fact even if it is not intuitively appealing. In the same way, copying and coding are complementary mechanisms of molecular change and there is no point in denying this fact by saying that one is a mechanism and the other a set of constraints.

This gives us an entirely new problem: what we need to understand is not the constraints that the organic codes have imposed on living systems, but the opportunities that they have created, the novelties that they have brought into existence, the role that they have played in macroevolution.

## 10.9 Major Steps in Macroevolution

The history of life is a 4,000-million-year-old story, and its greatest events are known as *Major Transitions* or, more simply, *Origins*. Here is a brief summary of them.

*The First Major Transition*

The origin of genes was due to the evolution of the first molecular machines that appeared spontaneously on the primitive Earth. These machines, the *bondmakers*, were molecules that could stick monomers together at random and produce polymers such as polypeptides, polynucleotides and polysaccharides. At a later stage, some bondmakers started joining monomers together no longer at random but in the order provided by a template. These bondmakers started making copies of other molecules and became *copymakers*. They were the machines that started copying nucleic acids, and gave origin to the molecules that we call genes.

*The Second Major Transition*
The origin of proteins was a much more complex affair, because proteins cannot be copied and yet the information to make them must come from molecules that can be copied, i.e., from nucleic acids. The problem is that there is no necessary link between nucleic acids and amino acids, and only the rules of a code can provide a bridge between them, which means that the evolution of protein synthesis had to go hand in hand with the evolution of the genetic code. The ancestral systems that developed the genetic code are collectively known as the *Common Ancestor*, but did not have the characteristics of modern cells.

*The Third Major Transition*
The origin of the first cells is associated with the appearance of the three primary kingdoms – Archaea, Bacteria and Eukarya – that evolved from the common ancestor in three different ways. But how? Here we should keep in mind that one the major priorities of the evolving cells was the ability to respond to the environment, and the mechanism of signal transduction was the key to that ability. The cells of the three domains have three distinct signalling systems, and this strongly suggests that the primary kingdoms arose from the combination of the universal genetic code with three distinct signal-transduction codes.

*The Fourth Major Transition*
The origin of eukaryotes took place in many stages, most of which were associated with new organic codes. The *splicing codes*, the *cytoskeleton codes*, the *compartment codes* and the *histone code* all introduced major novelties in the evolution of the eukaryotic cells and can rightly be regarded as the milestones of that evolution.

*The Fifth Major Transition*
The origin of embryos took also place in stages and required new organic codes. The codes of the body plan, in particular, were essential for this transition, and the fossil record clearly shows that all body plans were already present by the time of the Cambrian explosion and have never been changed ever since.

*The Sixth Major Transition*
The origin of mind is associated with the appearance of a nearly universal neural code, a code whose existence is supported by fact that the transformations of sense stimuli into neural sensations take place with virtually universal mechanisms. It is likely that many other neural codes appeared during the evolution of the brain, just as many organic codes appeared during the evolution of organic life. It must be underlined that *neural codes* were absolute novelties in the history of life because organic codes come from molecules, whereas neural codes come from action potentials and neuron firings.

*The Seventh Major Transition*
The origin of interpretation required processes of abduction that are more complex than coding, so it cannot be explained solely with the appearance of new codes. It is possible however that new codes did have a role also in this Major Transition.

*The Eight Major Transition*
The origin of language required the ability to make an extensive use of symbols, a breakthrough that apparently occurred only in our species and was associated with a third type of codes. After the organic codes and the neural codes, evolution gave also origin to the new world of the language codes.

This brief summary shows that all Major Transitions – except the first one – went hand in hand with the appearance of new codes, and this suggests that a close correspondence does exist between them, a conclusion that is reinforced by the concept that only codes have the potential to create absolute novelties.

## 10.10 Characteristics of the Organic Codes

The history of life has been 'punctuated' by the appearance of new organic codes and it has been deeply shaped by their characteristics. Six of them are particularly important.

1. *Discontinuity*
   The evolution of the individual rules of a code can take an extremely long time, but the origin of a new code corresponds to the appearance of a 'complete' set of rules and from a geological point of view that can be a sudden event. This is a new explanation of the discontinuities that paleontology has documented, and shows that natural selection and natural conventions had complementary roles. Natural conventions account for the discontinuities in the history of life whereas natural selection explains the gradual transformations that took place in between.
2. *Conservation*
   The genetic code appeared at the beginning of the history of life and has remained substantially the same ever since. The same applies to the other organic codes. Once in existence they have been conserved for billions of years, a fact that is often dismissed with the argument that somehow they have been *frozen*. In reality, all genetic components of a code are subject to constant mutation and selection, like all other genes, and only an *active* process that continuously restore them can ensure their conservation.
3. *Additivity*
   A new organic code has never abolished previous codes. The genetic code has not been removed by the signal transduction codes and neither of them has been supplanted by the splicing codes. A new code has always been *added* to the previous ones, which shows that new codes do not originate by the transformation

of previous codes. Once in existence, organic codes do not tend to change, and the origin of a new code is always the origin of an entirely new set of rules.

4. *Coexistence*
   Different living systems coexist whatever is the number of their codes. Eukaryotes did not remove prokaryotes, and multicellular systems did not remove free living single cells. Every functional arrangement of organic codes, in other words, represents an independent form of life that coexists with the others.
5. *Complexity*
   The structural complexity of some organisms can diminish in time, as many cases of simplification clearly show, but the complexity of the codes has never been lowered. Even the animals which lost or reduced a great number of parts in order to lead a parasitic life, have conserved the fundamental codes of animal life. The number of organic codes is therefore a new measure of biological complexity, and probably it is more fundamental than all the other parameters which have been proposed so far.
6. *Decreasing domains*
   The genetic code is present in all living creatures, but the other organic codes appeared in increasingly smaller groups. The greater the number of codes, the smaller the number of species that possess them. The ultimate result of this trend is that all known codes – organic, neural and language codes – exist in one species only.

## 10.11 Different Types of Codes

Protein synthesis is a manufacturing activity that is based on coding rules and we can say therefore that the genetic code is a *manufacturing code* because it is used for manufacturing purposes. Another example is splicing, the process that cuts away pieces of RNAs from primary transcripts and assembles the remaining pieces into messenger RNAs. Splicing is a manufacturing activity because it assembles specific objects from components and the splicing codes belong therefore to the category of the *manufacturing codes*.

There are however other organic codes that do not have a manufacturing function. Signal transduction, for example, is based on codes because it creates a correspondence between first and second messengers by means of receptor molecules that behave like adaptors, but it is not a manufacturing activity because the second messengers already exist in the cell and are not manufactured by the transduction process. Signal transduction is a *signalling* activity because its function is to create a specific signalling association between first and second messengers and the codes of signal transduction can therefore be referred to as *signalling codes*.

Other examples of signalling codes are the cytoskeleton codes and the compartment codes of the eukaryotic cells. In this cases too we recognize the existence of codes by the presence of adaptors and we realize that their function is to create specific associations between pre-existing objects, not to bring these objects into existence. Proteins, for example, are produced by a manufacturing activity but in order to arrange them in supramolecular structures like membrane, cytoskeleton and compartments we need other processes, and the existence of adaptors proves that many of these processes are based on codes. Manufacturing and signalling, in other words, have very different functions but they are both essential to life and represent therefore complementary processes.

Stefan Artmann (2009) has underlined that in addition to manufacturing and signalling there is in the cell a third activity that has a *controlling* or *regulating* function and that is also based on organic codes.

> ... we can observe, even in relatively simple living systems, a type of controlling semiosis that fulfils a metalingual function. The DNA sequence of an operon, i.e., the transcriptional unit for mRNA, contains a promoter that includes the operator. This is the site to which a repressor can bind in order to prevent that the protein-encoding parts (cistrons) of the operon are transcribed. The operator, thus, functions as a binary switch and determines whether cistrons are to be actually read.....Regulators, i.e., activators and repressors, play the role of adaptors in which the allosteric site and the active site do not bear a chemical resemblance. So it is reasonable to hypothesize a code in the switching logic of gene regulation.

The existence of multiple *regulatory codes* that overlap the genetic code has been proposed by various authors (Wheatheritt and Babu 2013). Stergachis et al., for example, have shown that ".... *genomes contain both a genetic code specifying amino acids and a regulatory code specifying transcription factor (TF) recognition sequences*" (Stergachis et al. 2013).

There are, in conclusion, three different types of organic codes in life – manufacturing, signalling and regulatory codes – and it is the combination of these codes that accounts for the complexity and for the evolutionary potential of the cells.

## 10.12 Unpredictable Features

The organic codes may give the impression of being deterministic rules that turn living systems into biological robots, but this is far from the truth. The key feature of the organic codes is the fact that they bring absolute novelties into existence and in so doing they produce objects that turn out to have totally *unpredictable* properties. This is a crucial point, and in order to illustrate it let us start from the case of those particular human artifacts that we call 'numbers'.

There is little doubt that numbers were generated by counting and that counting was favoured by natural selection because it had practical advantages. The process of counting, however, produces exclusively *natural* numbers, but then we have discovered *prime* numbers, *fractional* numbers, *rational* and *irrational* numbers, *real* and *imaginary* numbers, and in so doing we have brought to light an endless stream of mathematical theorems. All these *additional* entities were not produced by counting, and this is why some mathematicians say that natural numbers were *invented* by man but all other rules of mathematics had to be *discovered*, as if they had an existence of their own.

The world of mathematics was generated by the 'genetic' rule of counting and then it developed into an increasingly complex world full of additional properties. A world of codified objects is a world of *artifacts*, and it is only partially determined by the coding rules that generate the artifacts. In general, it turns out to have unpredictable 'rules of its own', rules that we call *epigenetic* because they were not present at the beginning and are brought to light only by processes of exploration and discovery.

This is what we actually find in living systems. In the world of proteins, for example, there is a universal mechanism in every cell that produces linear polypeptides from linear sequences of genes, but then the polypeptides fold themselves up into three-dimensional structures and take up forms that were not written in the genes. That generates a whole new world of objects, and living cells appear to engage in a veritable exploration of the potentialities of the protein universe.

Another outstanding example is the body-plan of animals. It is based on instructions that specify only three essential relationships between the cells of the body (up and down, back and front, left and right) and yet the number of morphological designs that can be built with them is virtually unlimited.

Yet another example is language, where the rules of grammar allow us to create endless combinations of words and generate the universe of oral and written human communication.

Language, mathematics, proteins and animals are very different worlds, but deep down there is something in common between them. They all have (1) a *genetic* algorithm that produces the objects of a potentially unlimited new world of artifacts (words, numbers, proteins and bodies) and (2) an exploratory procedure that brings into existence additional *epigenetic* properties of the new world that were not written in the coding rules and were not present at the beginning.

The organic codes, in conclusion, code for objects that not only are absolute novelties but also have unpredictable features. Far from being deterministic rules, they are the quintessential instruments of creativity and the higher their number the greater is the creative potential of a system. But they do not explain *everything*, far from it. They just account for the generative rules of life, not for the flesh and blood of history.

## 10.13 The Three Worlds of Life

The organic codes were the sole form of semiosis that existed on Earth in the first 3 billion years of evolution, but eventually two higher types of semiosis did appear. One evolved in nervous systems and gave animals the ability to *interpret* the world (*animal semiosis*). Interpretation is essentially what Peirce called an abduction, a process that is neither induction nor deduction but a 'generalization from limited

**Table 10.1** The three worlds of Popper correspond to three types of semiosis and to three different combinations of code-based mechanisms

| The Three Worlds of Life | | | |
|---|---|---|---|
| *Popper's Worlds* | *Type of Semiosis* | *Mechanisms* | *Codes* |
| WORLD 1 | Organic Semiosis | Coding | Organic codes |
| WORLD 2 | Animal Semiosis | Coding<br>Interpretation | Organic codes<br>Neural codes |
| WORLD 3 | Human Semiosis | Coding<br>Interpretation<br>Language | Organic codes<br>Neural codes<br>Language codes |

data' (Peirce 1906). More precisely, animals learned to interpret the world by using the two types of signs that Peirce called *icons* and *indexes*. They did not, however, exploit the third type of sign, the *symbols*. Only our species evolved that ability and developed a third type of semiosis that is based on language (*human semiosis*).

The evolution of life was characterized therefore by three great innovations: (1) the origin of organic semiosis, (2) the origin of animal semiosis, and (3) the origin of human semiosis. This fits nicely with the idea of the 'Three Worlds' proposed by Karl Popper (1972, 1979) because these worlds corresponds to three distinct types of semiosis, as illustrated in Table 10.1. This scheme shows that codes are fundamental components in all three worlds, but do not account for all types of semiosis. More precisely, it shows that coding is the sole mechanism of semiosis in the organic world, whereas animal semiosis is based on coding and interpretation, and human semiosis requires coding, interpretation and language.

## 10.14 Code Biology

Code biology is the study of all codes of life and this makes of it a well defined field of scientific research. The problem is: *what part of life do the codes account for*? Are they exceptional processes or fundamental components of the living systems? And if they are fundamental, why did the classical theories of biology ignore their existence?

The situation can perhaps be illustrated with an example from the history of science. In the eighteenth century, when physics was entirely based on Newton's laws, gravitation was the only known force in the universe and electricity was confined to the extravagant phenomena that were observed when some materials were rubbed together or when a kite was sent up in the sky during a storm to catch a lighting. It took a long time to change this view, but eventually it became clear that electricity too is a universal force of Nature and that the two forces can be studied separately because they account for very different aspects of reality.

In a similar way, in biology it has long been assumed that molecular copying is the sole fundamental mechanism of life, and in such a framework the genetic code can only be an extravagant singularity. Today we are beginning to realize that this is far from the truth. Molecular copying and molecular coding are both fundamental realities and the organic codes represent a whole new world of phenomena that so far has remained hidden from sight like electricity in the times of classical physics.

Code biology is the study of this new world, and it is, first and foremost, a new *experimental* field of research. But it is also a field that has outstanding theoretical consequences because it goes beyond the two official paradigms of modern biology. It goes beyond the Modern Synthesis with the idea that the great events of macroevolution were due to the appearance of new organic codes. And it goes beyond Systems Biology with the idea that the deep logic of life is not autopoiesis but codepoiesis.

From a practical point of view, it is important to underline that this new field of research has been established by the study of the organic codes, and this means that there are two distinct versions of Code Biology: a restricted version and a general one.

Code Biology in the strict sense is the study of all *organic* codes. Code Biology in the general sense is the study of *all* codes of life: organic codes, neural codes and language codes. This distinction is important because there are substantial differences between the two versions.

Code Biology in the strict sense is a fully scientific research field where the organic codes are studied with rigorous experimental methods and mathematical models. Code Biology in the general sense is, at the moment, a more speculative field because experiments on neural codes and language codes are much more difficult and in most cases we only have indirect evidence on them.

At the same time, we cannot concentrate exclusively on the organic codes. We must not lose sight of the larger picture formed by all codes of life, and this is why Code biology must study all of them, from the genetic code to the codes of culture, from the origin of life to the origin of language.

Code Biology, in conclusion, is a deeply interdisciplinary field that necessarily requires the contribution of different scholars: molecular biologists and cell biologists, evolutionary and developmental biologists, neuroscientists, cognitive scientists, ethologists, ecologists, mathematicians and computer scientists, linguists and philosophers.

## 10.15 There Was a Time

There was a time when *atoms* did not exist. They came into being within giant stars and were scattered all over the place when those stars exploded. There was a time when *molecules* did not exist. They originated from combinations of atoms on a variety of places such as comets and planets. There was a time when *molecular sequences* did not exist. They appeared when some molecules started joining other molecules together with chemical bonds and became *bondmakers*.

There was a time when *genes* did not exist. They came into being when some bondmakers joined nucleotides together in the order provided by a template and became *copymakers*, the molecular machines that started populating the Earth with copied molecules.

There was a time when *proteins* did not exist. They came into being when some supramolecular systems started joining amino acids together in the order provided by a sequence of nucleotides according to the rules of a code and became *codemakers*.

The spontaneous formation of molecules is fully described by the quantities of thermodynamics, but when a copymaker makes a copy of a template something new appears in the world: the sequence of the template becomes *information* for the copymaker. In a similar way, when a codemaker takes a chain of nucleotides to produce a chain of amino acids, something new comes into being: the second chain becomes the *meaning* of the first one. *It is the process of copying that creates information, and it is the process of coding that creates meaning*. Information and meaning appeared in the world when copymakers and codemakers came into existence and started populating the Earth with *manufactured* molecules, with *artifacts made by Nature*. And that was not just another step toward life. It was the origin of the very logic of life because it is the making of artifacts that divides life from chemistry. In a very fundamental sense, life is *artifact-making*. More precisely, it is *artifact-making by copying and coding*.

There was a time when *cells* did not exist. They came into being when a population of primitive systems that we call *common ancestor* evolved first a *genetic code* and then three distinct *signal transduction codes* that allowed them to respond to external signals and adapt to the environment.

There was a time when *embryos* did not exist. They came into being when some eukaryotic cells started producing clones of differentiating daughter cells that gave origin to symmetrical three-dimensional bodies by arranging themselves in space according to the coding rules of a body plan.

There was a time when *mind* did not exist. It came into being when some primitive animals evolved a population of intermediate neurons that started processing the incoming signals and generated the neural processes that we call feelings and instincts according to the rules of a nearly universal *neural code*.

There was a time when *language* did not exist. It came into being in one species that started giving birth to increasingly immature foetuses by displacing more and more stages of foetal development outside the uterus, thus forcing the growing creatures to develop an entirely new cognitive system, a new world of symbolic entities and imaginary creatures that turned them into *storytellers*.

There was a time when people did not know that organic codes exist in Nature. Now we are beginning to realize that most great events of macroevolution were associated with the origin of new organic codes. More than that. We are becoming aware of what really matters in living systems because it is an experimental fact that the organic codes are the great invariants of life, the sole entities that have been conserved for billions of years in evolution while everything else has been changed. And we trust that what is at the centre of life will also be, one day, at the centre of biology.

# References

Agalioti T, Chen G, Thanos D (2002) Deciphering the transcriptional histone acetylation code for a human gene. Cell 111(3):381–392
Alberts B, Johnson A, Lewis J, Raff M, Roberts K, Walter P (2007) Molecular biology of the cell, 5th edn. Garland, New York
Alerstram T, Hedenström A, Åkesson S (2003) Long-distance migration: evolution and determinants. Oikos 103:247–260
Amaral PP, Dinger ME, Mercer TR, Mattick JS (2008) The eukaryotic genome as an RNA machine. Science 319:1787–1789
Ambros V, Chen X (2007) The regulation of genes and genomes by small RNAs. Development 134:1635–1641
Anderson M, Deely J, Krampen M, Ransdell J, Sebeok TA, von Uexküll T (1984) A semiotic perspective on the sciences: steps toward a new paradigm. Semiotica 52(1/2):7–47
Andersson JO (2005) Lateral gene transfer in eukaryotes. Cell Mol Life Sci 62:1182–1197
Aristotle [384-322 BC] (1965) De Generatione Animalium. Translated by Platt A. Clarendon Press, Oxford
Arnellos A, Bruni LE, El-Hani CN, Collier J (2012) Anticipatory functions, digital-analog forms and biosemiotics. Biosemiotics 5:331–367
Artmann S (2007) Computing codes versus interpreting life: two alternative ways of synthesizing biological knowledge through semiotics. In: Barbieri M (ed) Introduction to biosemiotics: the new biological synthesis. Springer, Dordrecht, pp 209–233
Artmann S (2009) Basic semiosis as code-based control. Biosemiotics 2(1):31–38
Augustine (389 ad) (1963) De Doctrina Christiana. In: Green WM (ed) Sancti Augustini opera. CSEL 80. Hoelder-Pichler-Tempsky, Vienna
Avery OT, Macleod CM, McCarty M (1944) Studies on the chemical nature of the substance inducing transformation of pneumococcal types: induction of transformation by a desoxyribonucleic acid fraction isolated from pneumococcus Type III. J Exp Med 79:137–158
Baker M (2001) The atoms of language: the mind's hidden rules of grammar. Basic Books, New York
Ban N, Nissen P, Hansen J, Moore PB, Steitz TA (2000) The complete atomic structure of the large ribosomal subunit at 2.4 A resolution. Science 289:905–920
Barash Y, Calarco JA, Gao W, Pan Q, Wang X, Shai O, Blencow BJ, Frey BJ (2010) Deciphering the splicing code. Nature 465:53–59
Barbieri M (1981) The ribotype theory on the origin of life. J Theor Biol 91:545–601
Barbieri M (1985) The semantic theory of evolution. Harwood Academic Publishers, London/ New York

© Springer International Publishing Switzerland 2015 
M. Barbieri, *Code Biology*, DOI 10.1007/978-3-319-14535-8

Barbieri M (1998) The organic codes: the basic mechanism of macroevolution. Riv Biol Biol Forum 91:481–514

Barbieri M (2002) Has biosemiotics come of age? Semiotica 139(1/4):283–295

Barbieri M (2003) The organic codes: an introduction to semantic biology. Cambridge University Press, Cambridge

Barbieri M (2004) The definitions of information and meaning: two possible boundaries between physics and biology. Riv Biol Biol Forum 97:91–110

Barbieri M (2006a) Life and semiosis: the real nature of information and meaning. Semiotica 158(1/4):233–254

Barbieri M (2006b) Semantic biology and the mind-body problem: the theory of the conventional mind. Biol Theory 1(4):352–356

Barbieri M (2008) Biosemiotics: a new understanding of life. Naturwissenschaften 95:577–599

Barbieri M (2011) Origin and evolution of the brain. Biosemiotics 4(3):369–399

Barbieri M (2012) Codepoiesis – the deep logic of life. Biosemiotics 5(3):297–299

Barghoorn ES, Tyler SM (1965) Microfossils from the Gunflint chert. Science 147:563–577

Bartel DP (2009) MicroRNAs: target recognition and regulatory functions. Cell 136(2):215–233

Basañez G, Hardwick JM (2008) Unravelling the Bcl-2 apoptosis code with a simple model system. PLoS Biol 6(6):e154. doi:10.1371/journal.pbio.0060154

Bass BL (2002) RNA editing by adenosine deaminases that act on RNA. Annu Rev Biochem 71:817–846

Bateson W (1894) Materials for the study of variation. Macmillan and Co., London

Battail G (2007) Information theory and error-correcting codes in genetics and biological evolution. In: Barbieri M (ed) Introduction to biosemiotics. Springer, Dordrecht, pp 299–345

Battail G (2008) An outline of informational genetics. Morgan & Claypool Publishers, San Rafael

Battail G (2010) Heredity as an encoded communication process. IEEE Trans Inf Theory 56:678–687

Battail G (2014) Information and life. Springer Science + Business Media, Dordrecht

Bayascas JR, Castillo E, Saló E (1998) Platyhelminthes have a Hox code differentially activated during regeneration, with genes closely related to those of spiralian protostomes. Dev Genes Evol 208:467–473

Beadle G, Beadle M (1966) The language of life: an introduction to the science of genetics. Doubleday and Co, New York

Berger SL (2007) The complex language of chromatin regulation during transcription. Nature 447:407–412

Berridge M (1985) The molecular basis of communication within the cell. Sci Am 253:142–152

Bickerton D (1981) The roots of language. Karoma, Ann Arbour

Boeckx C (2006) Linguistic minimalism. Oxford University Press, New York

Bohr N (1933) Light and life. Nature 131:421–457

Boivin A, Vendrely R (1947) Sur le rôle possible deux acides nucleic dans la cellule vivant. Experientia 3:32–34

Bolk L (1926) Das Problem der Menschwerdung. Gustav Fischer, Jena

Bollenbach T, Kalin Vetsigian K, Kishony R (2007) Evolution and multilevel optimization of the genetic code. Genome Res 17:401–404

Boniolo G (2003) Biology without information. Hist Philos Life Sci 25:255–273

Bonnet C (1764) Contemplation de la Nature. Marc-Michel Rey, Amsterdam

Boutanaev AM, Mikhaylova LM, Nurminsky DI (2005) The pattern of chromosome folding in interphase is outlined by the linear gene density profile. Mol Cell Biol 18:8379–8386

Boveri TH (1904) Ergebnisse über die Konstitution der chromatischen Substanz des Zelkerns. Fisher, Jena

Boyd WC (1954) The proteins of immune reactions. In: Neurath H, Bayley K (eds) The proteins, Part 2, vol 2. Academic, New York, pp 756–844

Brachet J (1944) Embriologie Chimique. Masson et Cie, Paris

Brachet J (1946) Nucleic acids in the cell and the embryo. Symp Soc Exp Biol 1:213–215 and 222

Brandon MP, Hasselmo ME (2009) Sources of the spatial code within the hippocampus. Biol Rep 1:3–7

Brannon ME, Terrace HS (2000) Representation of the numerosities 1-9 by Rhesus Macaques (*Macaca mulatta*). J Exp Psychol 26(1):31–49

Brasier M, Cowie J, Taylor M (1994) Decision on the Precambrian-Cambrian boundary stratotype. Episodes 17:3–8

Brennicke A, Marchfelder A, Binder S (1999) RNA editing. FEMS Microbiol Rev 23(3):297–316

Brier S, Joslyn C (2013) What does it take to produce interpretation? Informational, Peircean and code-semiotic views on biosemiotics. Biosemiotics 6:143–159

Buckeridge MS, De Souza AP (2014) Breaking the "Glycomic Code" of cell wall polysaccharides may improve second-generation bioenergy production from biomass. Bioenergy Res. doi:10.1007/s12155-014-9460-6

Buckeridge MS, Crombie HJ, Mendes CJM, Reid JSG, Gidley MJ, Vieira CCJ (1997) A new family of oligosaccharides from the xyloglucan of *Hymenaea courbaril* L. (Leguminosae) cotyledons. Carbohydr Res 303(2):233–237

Buratti E, Baralle M, Baralle FE (2006) Defective splicing, disease and therapy: searching for master checkpoints in exon definition. Nucleic Acids Res 34:3494–3510

Burks AW (1970) Essays on cellular automata. University of Illinois Press, Urbana

Busch H, Smetana K (1970) The nucleolus. Academic, New York/London

Cantlon JF, Brannon EM (2006) Shared system for ordering small and large numbers in monkeys and humans. Psychol Sci 17(5):401–406

Carroll SB (1995) Homeotic genes and the evolution of arthropods and chordates. Nature 376:479–485

Carroll SB (2008) Evo-devo and an expanding evolutionary synthesis: a genetic theory of morphological evolution. Cell 134(1):25–36

Carroll SB, Grenier JK, Weatherbee SD (2001) From DNA to diversity: molecular genetics and the evolution of animal design. Blackwell, Malden

Cavalli G, Paro R (1998) Chromo-domain proteins: linking chromatin structure to epigenetic regulation. Curr Opin Cell Biol 10(3):354–360

Cech TR (1983) RNA splicing: three themes with variations. Cell 34:713–716

Cech TR (1986) RNA as an enzyme. Sci Am 255:64–75

Changeaux J-P (1983) L'Homme neuronal. Libraire Arthème Fayard, Paris

Charbon G, Breunig KD, Wattiez R, Vandenhaute J, Noël-Georis I (2004) Key role of Ser562/661 in Snf1-dependent regulation of Cat8p in *Saccharomyces cerevisiae* and *Kluyveromyces lactis*. Mol Cell Biol 24:4083–4091

Chargaff E (1963) Essays on nucleic acids. Elsevier, Amsterdam

Chomsky N (1957) Syntactic structures. Mouton, The Hague

Chomsky N (1965) Aspects of the theory of syntax. MIT Press, Cambridge, MA

Chomsky N (1975) The logical structure of linguistic theory. University of Chicago Press, Chicago

Chomsky N (1995) The minimalist program. MIT Press, Cambridge, MA

Chomsky N (2005) Three factors in language design. Linguist Inq 36:1–22

Churchland PS, Sejnowski TJ (1993) The computational brain. MIT Press, Cambridge, MA

Cloud P (1965) The significance of the Guntflint (Precambrian) microflora. Science 148:27–35

Collier J (2008) Information in biological systems. In: Adriaans P, van Benthem J (eds) Handbook of philosophy of science, volume 8, philosophy of information. Elsevier, Amsterdam

Conway Morris S (1993) The fossil record and the early evolution of the Metazoa. Nature 361:219–225

Conway Morris S (1998) The crucible of creation: the Burgess Shale and the rise of animals. Oxford University Press, Oxford

Cooper TA, Wan L, Dreyfuss G (2009) RNA and disease. Cell 136:777–793

Corballis MC (2013) The wandering rat: response to Suddendorf. Trends Cogn Sci 17(4):152

Cowley SJ (2007) How human infants deal with symbol grounding. Interact Stud 8:83–104

Cowley SJ (2011) Distributed language. John Benjamins, Amsterdam

Crick FHC (1957) The structure of nucleic acids and their role in protein synthesis. Biochemical Society symposium, 14. Cambridge University Press, Cambridge, pp 25–26
Crick FHC (1958) On protein synthesis. Symp Soc Exp Biol 12:138–163
Crick FHC (1966) Codon–anticodon pairing: the wobble hypothesis. J Mol Biol 19:548–555
Crick FHC (1994) The astonishing hypothesis: the scientific search for the soul. Scribner, New York
Csikszentmihalyi M (1996) Creativity: flow and the psychology of discovery and innovation. Harper Collins, New York
Cuvier G (1800–1805) Leçons d'Anatomie Comparée. 5 Volumes. Volumes 1–2 by Duméril, Paris. Volumes 3–5 by Duvernoy, Paris
Cuvier G (1812) Researches sur les ossements fossils des quadrupedès. Déterville, Paris
Cuvier G (1828) Le règne animal distribué d'apres son organization. Fortin, Paris
Dagan T, Artzy-Randrup Y, Martin W (2008) Modular networks and cumulative impact of lateral transfer in prokaryote genome evolution. Proc Natl Acad Sci U S A 105:10039–10044
Daily JM, Emery NJ, Clayton NS (2010) Avian Theory of Mind and counter espionage by food-caching western scrub-hays (Aphelocoma californica). Eur J Dev Psychol 7:17–37
Danchin A (2009) Bacteria as computers making computers. FEMS Microbiol Rev 33:3–26
Darwin C (1859) On the origin of species by means of natural selection, or preservation of favored races in the struggle for life. Murray, London
Dawkins R (1976) The selfish gene. Oxford University Press, Oxford
De Beule J, Hovig E, Benson M (2011) Introducing dynamics into the field of biosemiotics. Biosemiotics 4(1):5–24
de Saussure F (1916) Cours de linguistique générale. Payot, Paris
Deacon TW (1997) The symbolic species: the co-evolution of language and the brain. Norton, New York
Deacon TW (2006) Reciprocal linkage between self-organizing processes is sufficient for self-reproduction and evolvability. Biol Theory 1(2):136–149
Deacon TW (2007) Shannon-Boltzmann-Darwin: redefining information (Part I). Cogn Semiot 1(1):123–147
Deacon TW (2008) Shannon-Boltzmann-Darwin: redefining information (Part II). Cogn Semiot 1(2):168–198
DeHaan RL (1959) *Cardia bifida* and the development of pacemaker function in the early chicken heart. Dev Biol 1:586–602
Delbrück M (1971) Aristotle-totle-totle. In: Monod J, Borek E (eds) Of microbes and life. Columbia University Press, New York
Delbrück M (1986) Mind from matter? Blackwell Scientific Publications, Palo Alto
Des Marais DJ (1997) Isotopic evolution of the biogeochemical carbon cycle during the Proterozoic eon. Org Geochem 27:185–193
Descartes R (1637) *Discours de la Méthode*. L'Imprimerie de Ian Maire, Leiden, The Netherlands.
Dhir A, Buratti E, van Santen MA, Lührmann R, Baralle FE (2010) The intronic splicing code: multiple factors involved in ATM pseudoexon definition. EMBO J 29:749–760
Di Giulio M (2008) An extension of the coevolution theory of the origin of the genetic code. Biol Direct 3:1–37
Di Lorenzo PM (2000) The neural code for taste in the brain stem: response profiles. Physiol Behav 69:87–96
Dobzhansky T (1937) Genetics and the origin of species. Columbia University Press, New York
Donald M (1991) Origins of the modern mind. Harvard University Press, Cambridge, MA
Doolittle WF (1999) Phylogenetic classification and the universal tree. Science 284:2124–2129
Doolittle WF, Bapteste E (2007) Pattern pluralism and the tree of life hypothesis. Proc Natl Acad Sci U S A 104:2043–2049
Dounce AL (1952) Duplicating mechanism for peptide chain and nucleic acid synthesis. Enzymologia 15:251–258
Dounce AL (1953) Nucleic acid template hypothesis. Nature 172:541
Dover G, Flavell D (1982) Genome evolution. Oxford University Press, Oxford

Driesch H (1893) Zur Verlagerung der Blastomeren des Echinideneies. Anat Anz 8:348–357
Driesch H (1894) Analytische Theorie der Organischen Entwicklung. W. Engelmann, Leipzig
Dudai Y (1999) The smell of representations. Neuron 23:633–635
Edelman GM (1989) Neural Darwinism: the theory of neuronal group selection. Oxford University Press, New York
Eigen M (1971) Self-organization of matter and the evolution of biological macromolecules. Naturwissenschaften 58:465–523
Eigen M, Schuster P (1977) The hypercycle: a principle of natural self-organization. Naturwissenschaften 64:541–565
Elder D (1979) An epigenetic code. Differentiation 14:119–122
Farina A, Pieretti N (2014) Acoustic codes in action in a soundscape context. Biosemiotics 7(2):321–328
Favareau D (2007) The evolutionary history of biosemiotics. In: Barbieri M (ed) Introduction to biosemiotics. Springer, Dordrecht, pp 1–67
Fischer RA (1930) The genetical theory of natural selection. Oxford University Press, Oxford
Fischle W, Wang Y, Jacobs SA, Kim Y, Allis CD, Khorasanizadeh S (2003) Molecular basis for the discrimination of repressive methyl-lysine marks in histone H3 by Polycomb and HP1 chromodomains. Genes Dev 17(15):1870–1881
Fitch WM, Upper K (1987) The phylogeny of tRNA sequences provides evidence for ambiguity reduction in the origin of the genetic code. Cold Spring Harb Symp Quant Biol 52:759–767
Flames N, Pla R, Gelman DM, Rubenstein JLR, Puelles L, Marìn O (2007) Delineation of multiple subpallial progenitor domains by the combinatorial expression of transcriptional codes. J Neurosci 27(36):9682–9695
Florkin M (1974) Concepts of molecular biosemiotics and molecular evolution. In: Florkin M, Stotz EH (eds) Comprehensive biochemistry, vol 29A. Elsevier, Amsterdam, pp 1–124
Fodor J (1975) The language of thought. Thomas Crowell Co., New York
Fodor J (1983) The modularity of mind: an essay on faculty psychology. MIT Press, Cambridge, MA
Fox GE (2010) Origin and evolution of the ribosome. Cold Spring Harb Perspect Biol 2:a003483
Freeland SJ, Hurst LD (1998) The genetic code is one in a million. J Mol Evol 47:238–248
Frege G (1892) Über Sinn und Bedeutung. Zeitschrift fürPhilosophie und philosophische Kritik 100:25–50. English translation in Geach P, Black M (eds) Translations from the Philosophical Writings of Gottlob Frege. Blackwell, Oxford, 1952
Fu XD (2004) Towards a splicing code. Cell 119:736–738
Füllgrabe J, Hajji N, Joseph B (2010) Cracking the death code: apoptosis-related histone modifications. Cell Death Differ 17:1238–1243
Gabius H-J (2000) Biological information transfer beyond the genetic code: the sugar code. Naturwissenschaften 87:108–121
Gabius H-J (ed) (2009) The sugar code: fundamentals of glycosciences. Wiley-Blackwell, Weinheim
Gamow G (1954) Possible relation between deoxyribonucleic acid and protein structures. Nature 173:318
Garcia-Bellido A (1975) A genetic control of wing disc development in *Drosophila*. In: Cell patterning. Ciba Foundation symposium, 29. Elsevier, Amsterdam, pp 161–178
Garstang W (1922) The theory of recapitulation: a critical restatement of the biogenetic law. J Linn Soc London Zool 35:81–101
Gehring WJ (1996) The master control gene for morphogenesis and evolution of the eye. Genes Cells 1:11–15
Gell-Mann M (1994) The quark and the jaguar. W. H. Freeman, New York
Gilbert W (1986) The RNA world. Nature 319:618
Gilbert SF (2006) Developmental biology, 8th edn. Sinauer Associates Inc., Sunderland
Gimona M (2008) Protein linguistics and the modular code of the cytoskeleton. In: Barbieri M (ed) The codes of life: the rules of macroevolution. Springer, Dordrecht, pp 189–206

Glaessner MF (1983) The dawn of animal life: a biohistorical study. Cambridge University Press, Cambridge

Goethe JW (1790) Versuch der Metamorphose der Pflanzen zu Erklären. C. W. Ettinger, Gotha

Görlich D, Dittrich P (2013) Molecular codes in biological and chemical reaction networks. PLoS One 8(1):e54694. doi:10.1371/journal.pone.0054694

Görlich D, Artmann S, Dittrich P (2011) Cells as semantic systems. Biochim Biophys Acta 1810(10):914–923

Gould SJ (1977) Ontogeny and phylogeny. Harvard University Press, Cambridge, MA

Gould SJ (1989) Wonderful life: the Burgess Shale and the nature of history. Norton/Hutchinson Radius, London

Gould SJ (2002) The structure of evolutionary theory. Harvard University Press, Cambridge, MA

Gräff J, Mansuy IM (2008) Epigenetic codes in cognition and behavior. Behav Brain Res 192:70–87

Griffith PE (2001) Genetic information: a metaphor in search of a theory. Philos Sci 68:394–412

Griffith PE, Knight RD (1998) What is the developmental challenge? Philos Sci 65:276–288

Guerrier-Takada C, Altman S (1984) Catalytic activity of an RNA molecule prepared by transcription in vitro. Science 223:285–286

Guerrier-Takada C, Gardiner K, Marsh T, Pace N, Altman S (1983) The RNA moiety of ribonuclease P is the catalytic subunit of the enzyme. Cell 35:849–857

Hafting T, Fyhn M, Molden S, Moser MB, Moser EI (2005) Microstructure of a spatial map in the entorhinal cortex. Nature 436:801–806

Haig D, Hurst LD (1991) A quantitative measure of error minimization in the genetic code. J Mol Evol 33:412–417

Haldane JBS (1932) The causes of evolution. Longmans Green, London

Halder G, Callaerts P, Gehring WJ (1995) Induction of ectopic eyes by targeted expression of the *eyeless* gene in *Drosophila*. Science 267:1788–1792

Hallock RM, Di Lorenzo PM (2006) Temporal coding in the gustatory system. Neurosci Behav Rev 30:1145–1160

Hamburger V (1988) The heritage of experimental embryology: Hans Spemann and the organizer. Oxford University Press, Oxford/New York

Hamilton WD (1964) The genetical evolution of social behavior. J Theor Biol 7:1–32

Hansen JC, Tse C, Wolfe AP (1998) Structure and function of the core histone N-termini: more than meets the eye. Biochemistry 37:17637–17641

Harris R (1981) The language myth. Duckworth, London

Harrison RG (1918) Experiments on the development of the forelimb of *Amblystoma*, a self-differentiating equipotential system. J Exp Zool 25:413–461

Hauser MD, Chomsky N, Fitch WT (2002) The faculty of language: what is it, who has it, and how did it evolve? Science 298:1569–1579

Hennig W (1966) Phylogenetic systematic. University of Illinois Press, Urbana

Higgs PG, Pudritz RE (2007) From protoplanetary disks to prebiotic amino acids and the origin of the genetic code. In: Pudritz RE, Higgs PG, Stone J (eds) Planetary systems and the origins of life. Cambridge series in astrobiology, vol 3. Cambridge University Press, Cambridge

Hilschmann N, Barnikol HU, Barnikol-Watanabe S, Götz H, Kratzin H, Thinnes FP (2001) The immunoglobulin-like genetic predetermination of the brain: the protocadherins, blueprint of the neuronal network. Naturwissenschaften 88:2–12

Hoagland MB, Zamecnik PC, Stephenson ML (1957) Intermediate reactions in protein biosynthesis. Biochim Biophys Acta 24:215–216

Hoffmeyer J (1996) Signs of meaning in the universe. Indiana University Press, Bloomington

Hofmeyr J-H S (2007) The biochemical factory that autonomously fabricates itself: a systems-biological view of the living cell. In: Boogard F, Bruggeman F, Hofmeyr J, Westerhoff H (eds) Towards a philosophy of systems biology: Chapter 10. Elsevier, pp 217–242

Holland JA (1992) Adaptation in natural and artificial systems. MIT Press, Cambridge, MA

Hopfield JJ (1982) Neural networks and physical systems with emergent collective computational abilities. Proc Natl Acad Sci U S A 79:2554–2558

Hou Y-M, Schimmel P (1988) A simple structural feature is a major determinant of the identity of a transfer RNA. Nature 333:140–145
Hsiao C, Mohan S, Kalahar BK, Williams LD (2009) Peeling the onion: ribosomes are ancient molecular fossils. Mol Biol Evol 26:2415–2425
Hubel DH, Wiesel TN (1962) Receptive fields, binocular interaction and functional architecture in the cat's visual cortex. J Physiol 160:106–154
Hubel DH, Wiesel TN (1979) Brain mechanisms of vision. Sci Am 241(3):150–182
Hunt P, Whiting J, Nonchev S, Sham M-H, Marshall H, Graham A, Cook M, Alleman R, Rigby PW, Gulisano M (1991) The branchial *Hox* code and its implications for gene regulation, patterning of the nervous system and head evolution. Development 2:63–77
Huxley J (1942) Evolution: the modern synthesis. Allen and Unwin, London
Jacob F (1982) The possible and the actual. Pantheon Books, New York
Jacob F, Monod J (1961) Genetic regulatory mechanisms in the synthesis of proteins. J Mol Biol 3:318–356
James W (1890) The principles of psychology. Holt, New York. Reprinted in 1983 by Harvard University Press, Cambridge, MA
Jenuwein T, Allis CD (2001) Translating the histone code. Science 293:1074–1080
Jeppesen P (1997) Histone acetylation: a possible mechanism for the inheritance of cell memory at mitosis. BioEssays 19:67–74
Jessell TM (2000) Neuronal specification in the spinal cord: inductive signals and transcriptional codes. Nat Genet 1:20–29
Johannsen W (1909) Elemente der exacten Erblichkeitslehre. Gustav Fischer, Jena
Johnson-Laird PN (1983) Mental models. Harvard University Press, Cambridge, MA
Jukes TH, Osawa S (1990) The genetic code in mitochondria and chloroplasts. Experientia 46:1149–1157
Jukes TH, Osawa S (1993) Evolutionary changes in the genetic code. Comp Biochem Physiol 106B:489–494
Kalkowsky E (1908) Oolith und Stromatolith im norddeutschen Buntsandstein. Z Dtsch Geol Ges 60:68–125
Kandler O (1994) The early diversification of life. In: Bengston S (ed) Early life on earth: nobel symposium no. 84. Columbia University Press, New York, pp 152–160
Kant I (1781) Critique of Pure Reason. Translated by Smith NK (1929), Macmillan Press, London
Kelly SJ (1977) Studies of the developmental potential of 4- and 8-cell stage mouse blastomers. J Exp Zool 200:365–376
Kessel M, Gruss P (1991) Homeotic tansformation of murine vertebrae and concomitant alteration of *Hox* codes induced by retinoic acid. Cell 67:89–104
Khorana HG, Büchi H, Ghosh H, Gupta N, Jacob TM, Kössel H et al (1966) Polynucleotide synthesis and the genetic code. Cold Spring Harb Symp Quant Biol 31:39–49
Kim J, Daniel J, Espejo A, Lake A, Krishna M, Li X, Yi Z, Bedford MT (2006) Tudor, MBT and chromo domains gauge the degree of lysine methylation. EMBO Rep 7(4):397–403
Kirschner MW, Mitchison T (1986) Microtubule dynamics. Nature 324:621
Knoll AH (2003) Life on a young planet: the first three billion years of evolution on earth. Princeton University Press, Princeton
Kohonen T (1984) Self-organization and associative memory. Springer, New York
Kollmann J (1885) Das Überwintern von europäischen Frosch- und Tritonenlarven und die Umwaldung des mexikanischer Axolot. Verh Naturforsch Ges, Basel 7:387–398
Komander D, Rape M (2012) The ubiquitin code. Annu Rev Biochem 81:203–229
Kornberg RD, Lorch Y (1999) Twenty-five years of the nucleosome, fundamental particle of the eukaryote chromosome. Cell 98:285–294
Korzybski A (1996) Science and sanity. The International Non-Aristotelean Library Publishing Company, Lakeville
Krampen M (1981) Phytosemiotics. Semiotica 36(3/4):187–209
Kuhn T (1962) The structure of scientific revolutions. Chicago University Press, Chicago
Kühn S, Hofmeyr J-HS (2014) Is the "Histone Code" an organic code? Biosemiotics 7(2):203–222

Kull K (ed) (2001) Jakob von Uexküll: a paradigm for biology and semiotics. Semiotica, 134 (1/4). Mouton de Gruyter, Berlin

Kurland CG (1970) Ribosome structure and function emergent. Science 169:1171–1177

Laine RA (1977) The information-storing potential of the sugar code. In: Gabius H-J, Gabius S (eds) Glycosciences: status and perspectives. Chapman & Hall, London, pp 1–14

Lartigue C, Glass JI, Alperovich N, Pieper R, Parmar PP, Hutchison CA III, Smith HO, Venter JC (2007) Genome transplantation in bacteria: changing one species to another. Science 317: 632–638

Levenson JM, Sweatt JD (2005) Epigenetic mechanisms in memory formation. Nat Rev Neurosci 6(2):108–118

Levi-Montalcini R (1975) NGF: an uncharted route. In: Worden FG (ed) The neurosciences – paths of discoveries. MIT Press, Cambridge, MA

Levi-Montalcini R (1987) The nerve growth factor 35 years later. Science 237:1154–1162

Levin M (2014) Endogenous bioelectrical networks store non-genetic patterning information during development and regeneration. J Physiol 592(11):2295–2305

Lewis EB (1963) Genes and developmental pathways. Am Zool 3:33–56

Lewis EB (1978) A gene complex controlling segmentation in *Drosophila*. Nature 276:565–570

Libet B (1985) Unconscious cerebral initiative and the role of conscious will in voluntary action. Behav Brain Sci 8:529–566

Linde Medina M (2010) Two "EvoDevos". Biol Theory 5(1):7–11

Linnaeus C (1758) Systema Naturae per regna tria naturae, secundum classes, ordines, genera, species, cum characteribus, differentiis, synonymis, locis, 10th edn. Laurentius Salvius, Stockholm

Lorenz K (1971) Studies in animal and human behaviour. Harvard University Press, Cambridge, MA

Lotman J (1991) Universe of the mind: a semiotic theory of culture. Indiana University Press, Bloomington

Love N (2004) Cognition and the language myth. Lang Sci 26:525–544

Luisi PL, Varela FJ (1989) Self-replicating micelles – a chemical version of a minimal autopoietic system. Orig Life Evol Biosph 19:633–643

Lyko F, Paro R (1999) Chromosomal elements conferring epigenetic inheritance. BioEssays 21:824–832

Madison-Antenucci S, Grams J, Hajduk SL (2002) Editing machines: the complexities of trypanosome RNA editing. Cell 108:435–438

Mahner M, Bunge M (1997) Foundations of biophilosophy. Springer, Berlin

Maizels N, Weiner AM (1987) Peptide-specific ribosomes, genomic tags and the origin of the genetic code. Cold Spring Harb Symp Quant Biol 52:743–757

Malpighi M (1673) De formatione pulli in ovo. Royal Society, London

Maraldi NM (2008) A lipid-based code in nuclear signalling. In: Barbieri M (ed) The codes of life: the rules of macroevolution. Springer, Dordrecht, pp 207–221

Margueron R, Trojer P, Reinberg D (2005) The key to development: interpreting the histone code? Curr Opin Genet Dev 15:163–176

Margulis L (1970) Origin of eucaryotic cells. Yale University Press, New Haven

Markoš A (2002) Readers of the book of life: conceptualizing developmental evolutionary biology. Oxford University Press, Oxford

Markoš A, Grygar F, Hajnal L, Kleisner K, Kratochvfl Z, Neubauer Z (2009) Life as its own designer: Darwin's origin and western thought. Springer, Dordrecht

Marquard T, Pfaff SL (2001) Cracking the transcriptional code for cell specification in the neural tube. Cell 106:651–654

Marshall IN (1989) Consciousness and Bose-Einstein condensates. New Ideas Physiol 7:73–83

Maruta H, Greer K, Rosenbaum JL (1986) The acetylation of alpha-tubulin and its relationship to the assembly and disassembly of microtubules. J Cell Biol 103:571–579

Maslon L (1972) Wolf children and the problem of human nature. Monthly Review Press, New York

Matlin A, Clark F, Smith C (2005) Understanding alternative splicing: towards a cellular code. Nat Rev Mol Cell Biol 6:386–398

Mattick JS (2004) RNA regulation: a new genetics? Nat Rev Genet 5(4):316–323

Maturana HR, Varela FJ (1980) Autopoiesis and cognition: the realisation of the living. D. Reidel Publishing Company, Dordrecht

Maturana H, Varela F (1987) The tree of knowledge: the biological roots of human understanding. Shambhala, Boston

Maurer-Stroh S, Dickens NJ, Hughes-Davies L, Kouzarides T, Eisenhaber F, Ponting CP (2003) The Tudor domain 'Royal Family': tudor, plant agenet, chromo, PWWP and MBT domains. Trends Biochem Sci 28(2):69–74

Maynard Smith J (1986) The problems of biology. Oxford University Press, Oxford

Maynard Smith J (2000) The concept of information in biology. Philos Sci 67:177–194

Maynard Smith J, Szathmáry E (1995) The major transitions in evolution. Oxford University Press, Oxford

Mayr E (1942) Systematics and the origin of species. Columbia University Press, New York

Mayr E (1963) Animal species and evolution. Harvard University Press, Cambridge, MA

Mayr E (1982) The growth of biological thought. Harvard University Press, Cambridge, MA

McClintock B (1951) Chromosome organization and gene expression. Cold Spring Harb Symp Quant Biol 16:13–47

McClintock B (1956) Controlling elements and the gene. Cold Spring Harb Symp Quant Biol 21:197–216

McGinnis W, Levine M, Hafen E et al (1984) A conserved DNA sequence in homeotic genes of the *Drosophila* Antennapedia and Bithorax complexes. Nature 308:428–433

McLachlan J (1994) Medical embryology. Addison-Wesley, Amsterdam/New York

Melcher G (1974) Stereospecificity and the genetic code. J Mol Evol 3:121–141

Mellor J (2006) It takes a PHD to read the histone code. Cell 126(1):22–24

Mereschowsky C (1910) Theorie der Zwei Pflanzenarten als Grundlage der Symbiogenesis, einer neuen Lehre der Entstehung der Organismen. Biologisches Zentralblatt 30:278–303, 321–347, 353–367

Mesarovic MD (1968) Systems theory and biology. Springer, Berlin

Miller SL (1953) A production of amino acids under possible primitive earth conditions. Science 117:528–529

Miller RV (1998) Bacterial gene-swapping in nature. Sci Am 278(1):67–71

Mintz B (1962) Formation of genotypically mosaic embryos. Am Zool 2:432

Monod J (1970) Le Hasard et la Necessité. Seuil, Paris. English edition: (1971) Chance and Necessity. A Knopf, New York

Morgan TH (1915) The mechanism of Mendelian heredity. Henry Holt, New York

Morgan Lloyd C (1923) Emergent evolution. Williams and Norgate, London

Morowitz HJ (1992) Beginnings of cellular life. Yale University Press, New Haven

Mujtaba S, Zeng L, Zhou MM (2007) Structure and acetyl-lysine recognition of the bromodomain. Oncogene 26(37):5521–5527

Müller GB (2007) Evo-devo: extending the evolutionary synthesis. Nat Rev Genet 8(12):943–949

Murchison RI (1854) Siluria: the history of the oldest known rocks containing organic remains. John Murray, London

Nicolelis M (2011) Beyond boundaries: the new neuroscience of connecting brains with machines and how it will change our lives. Times Books, New York

Nicolelis M, Ribeiro S (2006) Seeking the neural code. Sci Am 295:70–77

Niesert U, Harnasch D, Bresch C (1981) Origin of life between Scylla and Charybdis. J Mol Evol 17:348–353

Niremberg M, Leder P (1964) RNA codewords and protein synthesis. Science 145:1399–1407

Niremberg M, Matthaei H (1961) The dependence of cell-free protein synthesis in *E. coli* upon naturally occurring or synthetic polyribonucleotides. Proc Natl Acad Sci U S A 47:1588–1602

Niremberg M, Caskey T, Marshal R, Brimacombe R, Kellogg D, Doctor B et al (1966) The RNA code and protein synthesis. Cold Spring Harb Symp Quant Biol 31:11–24

Nishimura S, Jones DS, Khorana HG (1965) The *in vitro* synthesis of a co-polypeptide containing two amino acids in alternating sequence dependent upon a DNA-like polymer containing two nucleotides in alternating sequence. J Mol Biol 13:302–324

Nitta I, Kamada Y, Noda H, Ueda T, Watanabe K (1998) Reconstitution of peptide bond formation. Science 281:666–669

Nomura M, Tissières A, Lengyel P (1974) Ribosomes. Cold Spring Harbor monograph series. Cold Spring Harbor Laboratory, New York

Nüsslein-Volhard C, Wieschaus E (1980) Mutations affecting segment number and polarity in *Drosophila*. Nature 287:795–801

O'Keefe J, Burgess N (1996) Geometric determinants of the place fields of hippocampal neurons. Nature 381:425–428

O'Keefe J, Burgess N (2005) Dual phase and rate coding in hippocampal place cells: theoretical significance and relationship to entorhinal grid cells. Hippocampus 15:853–866

Orgel LE (1973) The origins of life. Wiley, New York

Osawa S (1995) Evolution of the genetic code. Oxford University Press, Oxford/New York

Osborne LC, Palmer SE, Lisberger SG, Bialek W (2008) The neural basis for combinatorial coding in a cortical population response. J Neurosci 28(50):13522–13531

Owen R (1848) On the archetype and homologies of the vertebrate skeleton. John Van Voorst, London

Owen DJ, Ornaghi P, Yang J-C, Lowe N, Evans PR, Ballario P, Neuhaus D, Filetici P, Travers AA (2000) The structural basis for the recognition of acetylated histone H4 by the bromodomain of histone acetyltransferase Gcn5p. EMBO J 19(22):6141–6149

Panigrahi AK, Schnaufer A, Ernst NL, Wang B, Carmean N, Salavati R, Stuart K (2003) Identification of novel components of *Trypanosoma brucei* editosomes. RNA 9:484–492

Papoutsi M, de Zwart JA, Jansma JM, Pickering MJ, Bednar JA, Horwitz B (2009) From phonemes to articulatory codes: an fMRI study of the role of Broca's area in speech production. Cereb Cortex 19:2156–2165

Parker KA (1998) The continuity of Peirce's thought. Vanderbilt University Press, Nashville

Pattee HH (1968) The physical basis of coding and reliability in biological evolution. In: Waddington CH (ed) Toward a theoretical biology, vol 1. Edinburgh University Press, Edinburgh, pp 67–93

Pattee HH (1972) Laws and constraints, symbols and languages. In: Waddington CH (ed) Towards a theoretical biology, vol 4. Edinburgh University Press, Edinburgh, pp 248–258

Pattee HH (2001) The physics of symbols: bridging the epistemic cut. Bio Syst 60:5–21

Pattee HH (2008) Physical and functional conditions for symbols, codes and languages. Biosemiotics 1:147–168

Peirce CS (1906) The basis of pragmaticism. In: Hartshorne C, Weiss P (eds) The collected papers of Charles Sanders Peirce, vols I-VI (1931–1935). Harvard University Press, Cambridge, MA

Pelc SR, Weldon MGE (1966) Stereochemical relationship between coding triplets and amino-acids. Nature 209:868–870

Pepperberg IM, Carey S (2012) Grey parrot number acquisition: the inference of cardinal value from ordinal position on the numeral list. Cognition 125:219–232

Pertea M, Mount SM, Salzberg SL (2007) A computational survey of candidate exonic splicing enhancer motifs in the model plant *Arabidopsis thaliana*. BMC Bioinf 8:159

Peterson CL, Laniel M-A (2004) Histones and histone modifications. Curr Biol 14:546–551

Piaget J (1954) The construction of reality in the child. Basic Books, New York

Piaget J (1960) The child's conception of physical causality. Littlefield & Co., Paterson

Pigliucci M (2009) An extended synthesis for evolutionary biology: the year in evolutionary biology 2009. Ann N Y Acad Sci 1168:218–228

Pigliucci M, Müller GB (eds) (2010) Evolution – the extended synthesis. MIT Press, Cambridge, MA

Popper K (1972) Objective knowledge. Clarendon Press, Oxford

Popper K (1979) Three worlds. Mich Q Rev 18(1):1–23. Reprint of The Tanner Lecture on Human Values delivered by Karl Popper at The University of Michigan on April 7, 1978

Portier P (1918) Les symbiotes. Masson et Cie, Paris
Portmann A (1941) Die Tragzeiten der Primaten und die Dauer der Schwangerschaft beim Menschen: ein Problem der vergleichen Biologie. Rev Suisse Zool 48:511–518
Portmann A (1945) Die Ontogenese des Menschen als Problem der Evolutionsforschung. Verh Schweiz Naturf Ges 125:44–53
Posner R, Robering K, Sebeok TA (1997) Semiotik/semiotics: a handbook on the sign-theoretical foundations of nature and culture, vol 1. Walter de Gruyter, Berlin
Quiring R, Walldorf U, Kloter U, Gehring WJ (1994) Homology of the *eyeless* gene of Drosophila to the *Small eye* gene in mice and *Aniridia* in humans. Science 265:785–789
Raff RA (1996) The shape of life. The University of Chicago Press, Chicago/London
Ray A, van der Goes van Naters W, Shiraiwa T, Carlson JR (2006) Mechanisms of odor receptor gene choice in *Drosophila*. Neuron 53:353–369
Redies C, Takeichi M (1996) Cadherine in the developing central nervous system: an adhesive code for segmental and functional subdivisions. Dev Biol 180:413–423
Rensch B (1947) Evolution above the species level. Columbia University Press, New York
Roberts RB (1958) Microsomal particles and protein synthesis. Pergamon Press, Washington, DC
Rosen R (1958) A relational theory of biological systems. Bull Math Biophys 20:245–260
Rosen R (1991) Life itself: a comprehensive inquiry into the nature, origin, and fabrication of life. Columbia University Press, New York
Ruiz i Altaba A, Nguien V, Palma V (2003) The emergent design of the neural tube: prepattern, SHH morphogen and GLI code. Curr Opin Genet Dev 13:513–521
Rumelhart DE, McClelland JL (1986) Parallel distributed processing: explorations in the microstructure of cognition. MIT Press, Cambridge, MA
Ryan JF, Mazza ME, Pang K, Matus DQ, Baxevanis AD, Martindale MQ, Finnerty JR (2007) Pre-bilaterian origins of the Hox cluster and the Hox code: evidence from the sea anemome *Nematostella vectensis*. PLoS One 2:e153
Saint-Hilaire G (1818) Philosophie anatomique. J-B Baillière, Paris
Salvini-Plawen LV, Mayr E (1977) On the evolution of photoreceptors and eyes. In: Hecht MK, Steere WC, Wallace B (eds) Evolutionary biology, vol 10. Plenum, New York, pp 207–263
Sander K (1975) Pattern specification in the insect embryo. In: Cell patterning. Ciba Foundation symposium, 29. Elsevier, Amsterdam, pp 241–263
Sanders G (1970) Peirce's sixty-six signs? Trans Charles S Peirce Soc 6(1):3–16
Sarkar S (1996) Biological information: a skeptical look at some central dogmas of molecular biology. In: Sarkar S (ed) The philosophy and history of biology. Kluwer Academic Publishers, Dordrecht, pp 187–231
Sarkar S (2000) Information in genetics and developmental biology. Philos Sci 67:208–213
Saunders JW Jr (1982) Develpmental biology: patterns, problems, principles. Macmillan, New York
Schimmel P (1987) Aminoacyl tRNA synthetases: general scheme of structure-function relationship in the polypeptides and recognition of tRNAs. Annu Rev Biochem 56:125–158
Schimmel P, Giegé R, Moras D, Yokoyama S (1993) An operational RNA code for amino acids and possible relationship to genetic code. Proc Natl Acad Sci U S A 90:8763–8768
Schimper AFW (1883) Uber die Entwickelung der Chlorophyllkörner und Farbkorper. Bot Zeitung 41:105–114
Schopf JW (1978) The evolution of the earliest cells. Sci Am 239(3):84–103
Schopf JW (1993) Microfossils of the Early Archean Apex chert: new evidence of the antiquity of life. Science 260:640–646
Schopf JW (1999) Cradle of life: the discovery of earth's earliest fossils. Princeton University Press, Princeton
Schreiber SL, Bernstein BE (2002) Signaling network model of chromatin. Cell 111:771–778
Schrödinger E (1944) What is life? Cambridge University Press, Cambridge
Scott MP, Weiner AJ (1984) Structural relationships among genes that control development: sequence homology between the *Antennapedia*, *Ultrabithorax* and *fushi tarazu* loci in Drosophila. Proc Natl Acad Sci 81:4115–4119

Searle JR (1980) Minds, brains and programs. Behav Brain Sci 3:417–457
Searle JR (2002) Consciousness and language. Cambridge University Press, Cambridge
Sebeok TA (1963) Communication among social bees; porpoises and sonar; man and dolphin. Language 39:448–466
Sebeok TA (1972) Perspectives in Zoosemiotics. Mouton, The Hague
Sebeok TA (1991) A sign is just a sign. Indiana University Press, Bloomington
Sebeok TA (2001) Biosemiotics: its roots, proliferation, and prospects. In: Kull K (ed) Jakob von Uexküll: a paradigm for biology and semiotics. Semiotica 134(1/4):61–78
Sebeok TA, Danesi M (2000) The forms of meaning: modeling systems theory and semiotic analysis. Mouton de Gruyter, Berlin
Sebeok TA, Umiker-Sebeok J (eds) (1992) Biosemiotics. Mouton de Gruyter, Berlin
Segal E, Fondufe-Mittendorf Y, Chen L, Thåström A, Fiels Y, Moore IK, Wang JP, Widom J (2006) A genomic code for nucleosome positioning. Nature 442:772–778
Segré D, Ben Eli D, Deamer DW, Lancet D (2001) The lipid world. Orig Life Evol Biosph 31: 119–145
Seidel F (1952) Die Entwicklungspotenzen einer isolierten Blastomere des Zweizellenstadium im Säugetierei. Naturwissenschaften 39:355–356
Seilacher A (1992) Vendobionta and psammocarallia: lost constructions of Precambrian evolution. Geol Soc Lond J 149:607–613
Shannon CE (1948) A mathematical theory of communication. Bell Syst Tech J 27:379–424 and 623–656
Shapiro JA (2011) Evolution: a view from the 21st century. Prentice Hall, New York
Shapiro L, Colman DR (1999) The diversity of cadherins and implications for a synaptic adhesive code in the CNS. Neuron 23:427–430
Shattuck R (1981) The forbidden experiment: the story of the wild boy of Averyron. Washington Square Press, New York
Shen Q, Wang Y, Dimos JT, Fasano CA, Phoenix TN, Lemischka IR, Ivanova NB, Stifani S, Morrisey EE, Temple S (2006) The timing of cortical neurogenesis is encoded within lineages of individual progenitor cells. Nat Neurosci 9(6):743–751
Shepard RN, Metzler J (1971) Mental rotation of three-dimensional objects. Science 171:701–703
Shimizu M (1982) Molecular basis for the genetic code. J Mol Evol 18:297–303
Short TL (2007) Peirce's theory of signs. Cambridge University Press, Cambridge
Sibley CG, Ahlquist JE (1984) The phylogeny of the hominoid primates as indicated by DNA-DNA hybridization. J Mol Evol 20:2–15
Simpson GG (1944) Tempo and mode in evolution. Columbia University Press, New York
Slack JMW, Holland PWH, Graham CF (1993) The zootype and the phylotypic stage. Nature 361:490–492
Soja NN, Carey S, Spelke ES (1991) Ontological categories guide young children's inductions of word meaning: object terms and substance terms. Cognition 38:179–211
Solis AS, Shariat N, Patton JG (2008) Splicing fidelity, enhancers, and disease. Front Biosci 13:1926–1942
Sonea S (1988) The global organism: a new view of bacteria. The Sciences 28(4):38–45
Spemann H (1901) Über Korrelationen in der Entwicklung des Auges. Verh anat Ges Jena Verslg Bonn 15:61–79
Spemann H (1938) Embryonic development and induction. Yale University Press, New Haven
Spemann H, Mangold H (1924) Über Induktion von Embryonanlagen durch Implantation artfremder Organisatoren. Ark Micros Anat Entwmech 100:599–638
Sperry RW (1943) Visuomotor coordination in the newt (*Triturus viridescensis*) after regeneration of the optic nerve. J Comp Neurol 79:33–55
Sperry RW (1963) Chemoaffinity in the orderly growth of nerve fibers patterns and connections. Proc Natl Acad Sci U S A 50:703–710
Sperry RW (1968) Mental unity following surgical disconnection of the cerebral hemispheres. Harvey Lect 62:293–323

Sperry RW (1982) Some effects of disconnecting the cerebral hemispheres. Science 217:1223–1226

Speyer J, Lengyel P, Basilio C, Wahba A, Gardner R, Ochoa S (1963) Synthetic polynucleotides and the amino acid code. Cold Spring Harb Symp Quant Biol 28:559–567

Spiegelman S (1967) An in vitro analysis of a replicating molecule. Am Sci 55:3–68

Sporn MB, Roberts AB (1988) Peptide growth factors are multifunctional. Nature 332:217–219

Stapp HP (1993) Mind, matter and quantum mechanics. Springer, Berlin

Stent GS, Calendar R (1978) Molecular genetics. W.H. Freeman, San Francisco

Stergachis AB, Haugen E, Shafer A, Fu W, Vernot B, Reynolds A, Raubitschek A, Ziegler S, LeProust EM, Akey JM, Stamatoyannopoulos JA (2013) Exonic transcription factor binding directs codon choice and affects protein evolution. Science 342:1367–1372

Strahl BD, Allis D (2000) The language of covalent histone modifications. Nature 403:41–45

Suddendorf T (2013) Mental time travel: continuities and discontinuities. Trends Cogn Sci 17(4):151–152

Sutherland NS (1964) Visual discrimination in animals. Br Med Bull 20:54–59

Sutherland EW (1972) Studies on the mechanism of hormone action. Science 177:401–408

Sutton WS (1903) The chromosomes in heredity. Biol Bull 4:231–251

Swan LS, Goldberg LJ (2010) Biosymbols: symbols in life and mind. Biosemiotics 3(1):17–31

Szathmáry E (1999) Chemes, genes, memes: a revised classification of replicators. Lect Math Life Sci 26:1–10

Tarkowski AK (1961) Mouse chimeras developed from fused eggs. Nature 190:857–860

Tazi J, Bakkour N, Stamm S (2009) Alternative splicing and disease. Biochim Biophys Acta 1792:14–26

Tiné MAS, Silva CO, Lima DU, Carita NC, Buckeridge MS (2006) Fine structure of a mixed-oligomer storage xyloglucan from seeds of *Hymenaea courbaril*. Carbohydr Polym 66:444–454

Tomkins MG (1975) The metabolic code. Science 189:760–763

Trifonov EN (1987) Translation framing code and frame-monitoring mechanism as suggested by the analysis of mRNA and 16s rRNA nucleotide sequence. J Mol Biol 194:643–652

Trifonov EN (1989) The multiple codes of nucleotide sequences. Bull Math Biol 51:417–432

Trifonov EN (1996) Interfering contexts of regulatory sequence elements. *Cabios*, 12: 423–429.

Trifonov EN (1999) Elucidating sequence codes: three codes for evolution. Ann N Y Acad Sci 870:330–338

Tseng AS, Levin M (2013) Cracking the bioelectric code. Probing endogenous ionic controls of pattern formation. Commun Integr Biol 6(1):1–8

Tudge C (2000) The variety of life: a survey and a celebration of all the creatures that have ever lived. Oxford University Press, Oxford/New York

Turner BM (2000) Histone acetylation and an epigenetic code. BioEssays 22:836–845

Turner BM (2002) Cellular memory and the histone code. Cell 111:285–291

Turner BM (2007) Defining an epigenetic code. Nat Cell Biol 9:2–6

Tyler SM, Barghoorn ES (1954) Occurrence of structurally preserved plants in Precambrian rocks of the Canadian shield. Science 119:606–608

Van Helmont JB (1648) Ortus Medicinae. Elzevir, Amsterdam

Verhey KJ, Gaertig J (2007) The tubulin code. Cell Cycle 6(17):2152–2160

von Baer KE (1828) Über Entwickelungsgeschichte der Thiere: Beobachtung und Reflexion. Gebrüder Bornträger, Königsberg

von Baer KE (1866) De la découverte de larves qui se propagent. Bull Acad Impériale Sci St Petersbourg 9:63–137

von Bertalanffy L (1969) General system theory. George Braziller, New York

von Neumann J (1951) The general and logical theory of automata. In: Taub (1961), chap. 9, pp 288–328. Delivered at the Hixon Symposium, September 1948; first published 1951 as pages 1–41 in Jeffress A (ed) Cerebral mechanisms in behavior. Wiley, New York

von Neumann J (1956) Probabilistic logics and the synthesis of reliable organisms from unreliable components. In: Shannon CE, McCarthy J (eds) Automata studies. Princeton University Press, Princeton, pp 43–98

von Neumann J (1958) The computer and the brain. Yale University Press, New Haven
von Neumann J (1966) The theory of self-reproducing automata. Edited and completed by Burks A, University of Illinois Press, Urbana
von Uexküll J (1909) Umwelt und Innenwelt der Tiere. Julius Springer, Berlin
von Uexküll J (1928) Theoretische Biologie 2te Auflage. Julius Springer, Berlin
Wächtershäuser G (1997) The origin of life and its methodological challenge. J Theor Biol 187:483–494
Wallin JE (1927) Symbionticism and the origin of species. Williams and Wilkins, Baltimore
Wang Z, Burge C (2008) Splicing regulation: from a part list of regulatory elements to an integrated splicing code. RNA 14:802–813
Wang GS, Cooper TA (2007) Splicing in disease: disruption of the splicing code and the decoding machinery. Nat Rev Genet 8:749–761
Wang GG, Cai L, Pasillas MP, Kamps MP (2007) NUP98–NSD1 links H3K36 methylation to Hox-A gene activation and leukaemogenesis. Nat Cell Biol 9:804–812
Watson JD, Crick FHC (1953) Molecular structure of nucleic acids: a structure for deoxyribose nucleic acid. Nature 171:737–738. Genetical implications of the structure of deoxyribose nucleic acid. Nature 171:964–967
Webster DR, Wehland J, Weber K, Borisy GG (1990) Detyrosination of alpha tubulin does not stabilize microtubules in vivo. J Cell Biol 111:113–122
Wheatheritt RJ, Babu MM (2013) The hidden codes that shape protein evolution. Science 342:1325–1326
Wiener N (1948) Cybernetics: or control and communication in the animal and the machine. Hermann, Paris
Wigner E (1964) Events, laws, and invariance principles. Wigner's Nobel Lecture, Stockholm, 10 December 1963. Reprinted in Science 145:995–999
Williams GC (1966) Adaptation and natural selection. Princeton University Press, Princeton
Wimberly BT, Brodersen DE, Clemons WM Jr, Morgan-Warren RJ, Carter AP, Vonrhein C, Hartsch T, Ramakrishnan V (2000) Structure of the 30S ribosomal subunit. Nature 407: 327–339
Winterburn PJ, Phelps CF (1972) The significance of glycosylated proteins. Nature 236:147–151
Woese CR (1965) Order in the genetic code. Proc Natl Acad Sci U S A 54:71–75
Woese CR (1987) Bacterial evolution. Microbiol Rev 51:221–271
Woese CR (2000) Interpreting the universal phylogenetic tree. Proc Natl Acad Sci U S A 97:8392–8396
Woese CR (2002) On the evolution of cells. Proc Natl Acad Sci U S A 99:8742–8747
Woese CR, Fox GE (1977) Phylogenetic structure of the prokaryotic domain: the primary kingdoms. Proc Natl Acad Sci U S A 74:5088–5090
Woese CR, Dugre DH, Saxinger WC, Dugre SA (1966) The molecular basis for the genetic code. Proc Natl Acad Sci U S A 55:966–974
Woese CR, Kandler O, Wheelis ML (1990) Towards a natural system of organisms: proposal for the domains Archaea, Bacteria and Eukarya. Proc Natl Acad Sci U S A 87:4576–4579
Wolf YI, Koonin EV (2007) On the origin of the translation system and the genetic code in the RNA world by means of natural selection, exaptation, and subfunctionalization. Biol Direct 2:14
Wolfe AP, Hayes JJ (1999) Chromatin disruption and modification. Nucleic Acid Res 27:711–720
Wong JTF (1981) Coevolution of genetic code and amino acid biosynthesis. TIBS 33–36
Wong JT, Bronskill PM (1979) Inadequacy of prebiotic synthesis as origin of proteinous amino acids. J Mol Evol 13:115–125
Wright S (1931) Evolution in Mendelian populations. Genetics 16:97–159
Wu J, Grunstein M (2000) 25 years after the nucleosome model: chromatin modifications. Trends Biochem Sci 25(12):619–623
Xin L, Zhou G-L, Song W, Wu X-S, Wei G-H, Hao D-L, Lv X, Liu D-P, Liang C-C (2007) Exploring cellular memory molecules marking competent and active transcriptions. BioMed Cent Mol Biol 8:31–40

Yartsev MM, Ulanovsky N (2013) Representation of three-dimensional space in the hippocampus of flying bats. Science 340:367–372

Yockey HP (1974) An application of information theory to the Central Dogma and the sequence hypothesis. J Theor Biol 46:369–406

Yockey HP (2000) Origin of life on earth and Shannon's theory of communication. Comput Chem 24:105–123

Yockey HP (2005) Information theory, evolution, and the origin of life. Cambridge University Press, Cambridge

Yockey HP, Platzman RL, Quastler H (eds) (1958) Symposium on information theory in biology. Pergamon Press, New York/London

Yuan GC, Liu YJ, Dion MF, Slack MD, Wu LF, Altschuler SJ, Rando OJ (2005) Genome-scale identification of nucleosome positions in *S. cerevisiae*. Science 309:626–630

Zhu B, Zheng Y, Pham A-D, Mandal SS, Erdjument-Bromage H, Tempst P, Reinberg D (2005) Monoubiquitination of human histone H2B: the factors involved and their roles in *HOX* gene regulation. Mol Cell 20(4):601–611

Zohar D (1990) The quantum self. William Morrow, New York

# Author Index


**A**
Agalioti, T., 46
Ahlquist, J.E., 136
Alberts, B., xiii, 43, 44, 95, 99
Alerstram, T., 145
Allis, D., xi, 46
Altman, S., 59
Amaral, P.P., 83
Ambros, V., 83
Anderson, M., 151, 158, 166
Andersson, J.O., 89
Aristotle, 93, 94, 97, 109, 130, 146, 155, 165
Arnellos, A., 167
Artmann, S., 26, 168, 184
Avery, O.T., 36

**B**
Babu, M.M., 52, 184
Baker, M., 132
Ban, N., 65
Bapteste, E., 89
Baralle, F.E., 40
Baralle, M., 40
Barash, Y., xi, 41
Barbieri, M., xi, xii, xv, 11, 12, 23, 25, 26, 32, 40, 44, 60–62, 64, 72, 73, 87, 88, 110, 111, 124, 132, 161, 162, 164, 176, 178
Barghoorn, E.S., 76
Bartel, D.P., 83
Basañez, G., xi, 121
Bass, B.L., 83
Bateson, W., 101
Battail, G., 7
Bayascas, J.R., 107
Beadle, G., 35, 152
Beadle, M, 35, 152
Berger, S.L., 46
Bernstein, B.E., 46
Berridge, M., xiii, 43
Bickerton, D., 140
Boeckx, C., 132
Bohr, N., 153
Boivin, A., 36
Bolk, L., 133, 134
Bollenbach, T., 38
Boniolo, G., 10
Bonnet, C., 94, 95
Boutanaev, A.M., 52
Boveri, T.H., 84
Boyd, W.C., 50
Brachet, J., 36, 37
Brandon, M.P., 117
Brannon, E.M., 145
Brasier, M., 105
Brennicke, A., 83
Brier, S., 167
Bronskill, P.M., 67
Bruni, L.E., 167
Buckeridge, M.S., xi, 51
Bunge, M., 10
Buratti, E., 40
Burge, C., xi, 40
Burgess, N., 117
Burks, A.W., 70
Busch, H., 62


© Springer International Publishing Switzerland 2015
M. Barbieri, *Code Biology*, DOI 10.1007/978-3-319-14535-8

**C**
Calendar, R., 20
Cantlon, J.F., 145
Carey, S., 145
Carroll, S.B., 102, 103, 171, 177
Cavalli, G., 48
Cech, T.R., 59
Changeaux, J.-P., 119
Charbon, G., 45
Chargaff, E., 10
Chen, X., 83
Chomsky, N., 130–132
Churchland, P.S., 122
Cloud, P., 77
Collier, J., 15
Colman, D.R., xi, 120
Conway Morris, S., 105–107
Cooper, T.A., 40
Corballis, M.C., 145
Cowley, S.J., 141, 144, 145
Crick, F.H.C., 5, 6, 36–38, 45, 52, 122
Csikszentmihalyi, M., 168
Cuvier, G., 95, 96, 104

**D**
Dagan, T., 80, 89
Daily, J.M., 145
Danchin, A., 6
Danesi, M., 126
D'Arcy Thomson, 5
Darwin, C., 88, 104, 138, 155, 159, 160, 171, 176, 177
Dawkins, R., 177
De Beule, J., xii, 168
de Saussure, F., 28, 131, 146, 157
De Souza, A.P., xi, 51
Deacon, T.W., 155
DeHaan, R.L., 136, 137
Delbrück, M., 94, 116
Des Marais, D.J., 105
Descartes, R., 4, 16
Dhir, A., xi, 41
Di Giulio, M., 67
Di Lorenzo, P.M., 117
Dittrich, P., xii, 168
Dobzhansky, T., 171
Donald, M., 146
Doolittle, W.F., 63, 89
Dounce, A.L., 36
Dover, G., 84
Driesch, H., 100, 160
Dudai, Y., 117

**E**
Edelman, G.M., 122
Eigen, M., 59, 60
Elder, D., xi, 107
El-Hani, C.N., 167
Emmeche, C., 161

**F**
Farina, A., xi, 168
Favareau, D., 32, 155, 158
Fischer, R.A., 171
Fischle, W., 48
Fitch, W.M., 68
Flames, N., 121
Flavell, D., 84
Florkin, M., 151, 153, 158
Fodor, J., 121
Fox, G.E., 66, 69, 70, 80, 82, 90
Freeland, S.J., 38, 71
Frege, G., 164, 165
Fu, X.D., xi, 40, 41
Füllgrabe, J., xi, 121

**G**
Gabius, H.-J., xi, 50
Gadamer, H.G., 159
Gaertig, J., xi, 49
Gamow, G., 39
Garcia-Bellido, A., 102
Garstang, W., 133
Gehring, W.J, 103
Gell-Mann, M., 153
Gilbert, S.F., 101, 102, 108
Gilbert, W., 59
Gimona, M., xi
Glaessner, M.F., 105
Goethe, J.W., 5, 104, 160
Goldberg, L.J., 41
Görlich, D., xii, 168
Gould, S.J., 79, 106, 134, 171, 177
Gräff, J., 46
Graham, C.F., 104
Griffith, F., 36
Grunstein, M., 46
Gruss, P., 107
Guerrier-Takada, C., 59

**H**
Hafting, T., 117
Haig, D., 38, 71
Haldane, J.B.S., 171

Halder, G., 103
Hallock, R.M., 117
Hamburger, V., 98
Hamilton, W.D., 176
Hansen, J.C., 46
Hardwick, J.M., xi, 121
Harris, R., 146
Harrison, R.G., 101
Hasselmo, M.E., 117
Hauser, M.D., 130
Hayes, J.J., 46
Heidegger, M., 159, 160
Hennig, W., 54, 104
Higgs, P.G., 14, 67
Hilschmann, N., 120
Hoagland, M.B., 37
Hoffmeyer, J., 158, 159, 161, 162
Hofmeyr, J.-H.S., xi, 47, 48, 168
Holland, P., 104
Hopfield, J.J., 122, 125
Hou, Y.-M., xiii, 39
Hsiao, C., 65
Hubel, D.H., 115, 116
Hunt, P., 107
Hurst, L.D., 38, 71
Huxley, J., 171

**J**
Jacob, F., 37, 81, 138
James, W., 113
Jenuwein, T., xi, 46
Jeppesen, P., 46
Jessell, T.M., 121
Johannsen, W., 5, 73
Johnson-Laird, P.N., 121
Joslyn, C., 167
Jukes, T.H., 39

**K**
Kalkowsky, E., 76
Kandler, E., 80
Kant, I., 138
Kauffman, S., 5, 160
Kelly, S.J., 100
Kessel, M., 107
Khorana, H.G., 37, 53, 152
Kim, J., 48
Kirschner, M.W., 48
Knight, R.D., 10
Knoll, A.H., xiv, 76, 77, 79, 81
Kohonen, T., 125
Kollmann, J., 133, 134
Komander, D., xi
Koonin, E.V., 66
Kornberg, R.D., 46
Korzybski, A., 16
Krampen, M., 158
Kühn, S., xi, 47, 48, 168
Kuhn, T., 168
Kull, K., 158–161
Kurland, C.G., 69

**L**
Laine, R.A., 50
Laniel, M.-A., 46
Lartigue, C., 6
Leder, P., 37
Leeuwenhoek, 94
Leibnitz, 94
Levenson, J.M., 47
Levi-Montalcini, R., 119
Levine, M., xii
Lewis, E.B., 101, 102
Libet, B., 115
Linde Medina, M., 171
Linnaeus, C., 104
Lorch, Y., 46
Lorenz, K., 138, 142
Lotman, J., 126, 130, 131
Love, N., 146
Luisi, P.L., 64
Lyko, F., 46

**M**
Madison-Antenucci, S., 83
Mahner, M., 10
Maizels, N., 68
Malpighi, M., 94
Mangold, H., 98
Mansuy, I.M., 46
Maraldi, N.M., xi
Margueron, R., 46
Margulis, L., 79, 80
Markoš, A., 159–162
Marquard, T., 121
Marshall, I.N., 122
Maruta, H., 48
Maslon, L., 140
Matlin, A., xi, 40, 41
Matthaei, H., 37, 152
Mattick, J.S., 83

Maturana, H., 146
Maurer-Stroh, S., 48
Maynard-Smith, J., 5, 16, 60, 109, 110, 177
Mayr, E., 3, 9, 94, 103, 154, 171, 177
McClelland, J.L., 122, 125
McClintock, B., 84
McGinnis, W., 103
McLachlan, J., 100
Melcher, G., 39
Mellor, J., 48
Mendel, 5, 171, 176
Mereschowsky, C., 79
Mesarovic, M.D., 171
Metzler, J., 121
Miller, R.V., 89
Miller, S.L., 57
Mintz, B., 100
Mitchison, T., 48
Monod, J., 37, 40, 81, 94
Morgan Lloyd, C., 122
Morgan, T.H., 84
Morowitz, H.J., 64
Mujtaba, S., 47
Müller, G.B., 171, 177
Murchison, R.I., 105

**N**

Nicolelis, M., 118, 124
Niesert, U., 60
Niremberg, M., 37, 53, 152
Nishimura, S., 37
Nitta, I., 61, 65
Nomura, M., 37, 65
Nüsslein-Volhard, C., 102

**O**

O'Keefe, J., 117
Orgel, L.E., 60
Osawa, S., 39, 68
Osborne, L.C., 121
Ottolenghi, C., 168
Owen, D.J., 47
Owen, R., 104

**P**

Panigrahi, A.K., 83
Papoutsi, M., 117
Parker, K.A., 155
Paro, R., 46, 48
Pattee, H.H., 21, 22, 152–155
Peirce, C.S., xvi, 28, 125, 132, 143, 152, 155–158, 164, 166–168, 185, 186
Pelc, S.R., 39
Pepperberg, I.M., 145
Pertea, M., xi, 41
Peterson, C.L., 46
Pfaff, S.L., 121
Phelps, C.F., 50
Piaget, J., 138–140
Pieretti, N., xi
Pigliucci, M., 171, 177
Polanyi, M., 153
Popper, K., 186
Portier, P., 79
Portmann, A., 134
Posner, R., 158
Prigogine, I., 5
Pudritz, R.E., 14, 67

**Q**

Quiring, R., 103

**R**

Raff, R.A., 101
Rape, M., xi
Ray, A., 117
Redies, C., xi, 120
Rensch, B., 171
Ribeiro, S., 118, 124
Roberts, A.B., 44
Roberts, R.B., 37
Rosen, R., vii, ix, 171
Ruiz i Altaba, A., 121
Rumelhart, D.E., 122, 125
Ryan, J.F., 107

**S**

Saint-Hilaire, G., 104, 133
Salvini-Plawen, L.V., 103
Sander, K., 102
Sanders, G., 155
Sarkar, S., 10
Saunders, J.W. Jr., 101
Schimmel, P., xiii, 39, 70
Schimper, A.F.W., 79
Schopf, J.W., xiv, 75, 76, 78, 81
Schreiber, S.L., 46
Schrödinger, E., 20, 153
Schuster, P., 60

Scott, M.P., 103
Searle, J.R., 122
Sebeok, T.A., xvi, 126, 130–132, 151, 156–162, 166, 167
Segal, E., 52
Segré, D., 64
Seidel, F., 100
Seilacher, A., 105, 107
Sejnowski, T.J., 122
Shannon, C.E., 6, 7, 9, 21, 34, 52, 154, 155
Shapiro, J.A., 85, 120
Shapiro, L., xi, 120
Shattuck, R., 140
Shen, Q., 121
Shepard, R.N., 121
Shimizu, M., 39
Short, T.L., 155
Sibley, C.G., 136
Simpson, G.G., 171
Slack, J.M.W., 104, 105
Smetana, K., 62
Soja, N.N., 140
Solis, A.S., 40
Sonea, S., 89
Spallanzani, 94
Spemann, H., 97–99
Sperry, R.W., 120, 141
Speyer, J., 37, 152
Spiegelman, S., 59
Sporn, M.B., 44
Stahl G. E., 4
Stapp, H.P., 122
Stent, G.S., 20
Stergachis, A.B., 52, 184
Strahl, B.D., xi, 46
Suddendorf, T., 145
Sutherland, E.W., 43
Sutherland, N.S., 145
Sutton, W.S., 84
Swan, L.S., 41, 168
Sweatt, J.D., 47
Szathmáry, E., 60, 64, 177

**T**
Takeichi, M., xi, 120
TØnnessen, M., 168
Tarkowski, A.K., 100
Tazi, J., 40
Terrace, H.S., 145
Thom R., 5, 160
Tiné, M., 51
Tomkins, M.G., xi, 41, 42
Trifonov, E.N., xi, 52, 53
Tseng, A.S., xii
Tudge, C., 106
Turner, B.M., xi, 46
Tyler, S.M., 76

**U**
Ulanovsky, N., 154
Umiker-Sebeok, J., 151, 166
Upper, K., 68

**V**
van Helmont, J.B., 10
Varela, F.J., 64, 146, 171
Vendrely, R., 36
Verhey, K.J., xi, 49
von Baer, K.E., 95–97, 104, 133, 160
von Bertalanffy, L., 171
von Neumann, J., 21, 22, 152–154, 171, 173–176
von Uexküll, J., 126, 130, 131, 156–158, 160, 161, 167

**W**
Wächtershäuser, G., 10
Waddington, C.H., 152, 160
Walcott, C., 76
Wallin, J.E., 79
Wang, G.G., 46
Wang, G.S., 40
Wang, Z., xi, 40
Watson, J.D., 5, 6, 36, 45, 52
Webster, D.R., 48
Weiner, A.M., 68
Weldon, M.G.E., 39
Wheatheritt, R.J., 52, 184
Wiener, N., 171
Wieschaus, E., 102
Wiesel, T.N., 115, 116
Wigner, E., 21, 154
Williams, G.C., 171, 176, 177
Wills, P., 168
Wimberly, B.T., 65
Winterburn, P.J., 50
Woese, C.R., xiv, 38, 60, 63, 70, 71, 80, 82, 90

Wolf, Y.I., 66
Wolfe, A.P., 46
Wong, J.T.F., 14, 67
Wright, S., 171, 177
Wu, J., 46

**X**
Xin, L., 46

**Y**
Yartsev, M.M., 145
Yockey, H.P., 9, 12
Yuan, G.C., 52

**Z**
Zhu, B., 46
Zohar, D., 122

# Subject Index


**A**
Abduction, 125, 126, 155, 164, 182, 185
Absolute novelties, 123, 124, 159, 164, 178, 181, 182, 184, 185
Accessory proteins, 87
Acetylation, 45–46, 48
Acetylcholine, 44
Acidophiles, 82
Acoustic codes, xi
Action potential(s), 46–47, 112, 113, 181
Activating enzymes, 36
Activation, 43, 44, 46, 47, 102, 119, 120, 140, 142, 143, 184
Adaptations, 54, 102, 138
Adaptive evolution, 176
Adaptors, xii, xiii, 12, 35, 37, 39–41, 43, 44, 47–50, 53, 83, 86, 87, 108, 109, 142, 146–147, 183–184
Adhesive code, xi, 120
Adrenaline, 44
Aerobes, 78, 79
Age of manifest life, 79
Age of the bluebacteria, 77
Age of the cells, 75
Age of the protista, 78–79
Agglutinins, 50
Algorithm, 16, 110, 160, 185
Alternative splicing, 40
Ambiguity, 68–70, 117
Amino acids, xiii, 5, 9, 12, 14, 24, 26, 31–33, 36–39, 41, 42, 47, 48, 50, 53, 61, 65–72, 83, 88, 102, 111, 124, 142, 162, 167, 174, 181, 184, 188
Amino acid starvation, 42
Aminoacyl-tRNA-synthetases, xiii, 36, 39, 53, 68, 70, 72
Amnion, 100
Amphibians, 96, 99, 116
Anaerobes, 77, 78
Analog, 9, 12, 42, 96, 129, 136–137, 167
Analog signals, 7
Analogy, 139
Anatomical designs, 96
Ancestors, xiv, 38, 61, 66, 68–70, 72, 75, 80–82, 90, 91, 103, 104, 106, 107, 126, 129–131, 134–137, 142, 144, 145, 175, 181, 188
Ancestral genetic code, 68–70
Ancestral ribosomes, 65
Anchoring-RNAs, 66
Ancient genetic code, 68–71
Anima irrationalis, 146
Animal phyla, 95–97, 102–106, 109
Animal semiosis, 156, 157, 163–164, 166, 185, 186
Anima rationalis, 146
Annelid worms, xi, 107
Antennapedia, 101
Anthroposemiosis, 166
Anticodon, 37–39, 65
Aperiodic crystal, 20
Apoptosis, 119–121
Apoptosis code, xi, 119–121
Apparatus of protein synthesis, 33, 36–37, 59, 65, 66, 68, 70, 71, 146, 165, 174
Arbitrariness, 35, 39–40, 48, 143
Arbitrary rules, xii, 32, 35, 40, 87
Archaea, xiv, xvi, 72, 80, 81
Archaebacteria, xiv, 80, 82
*Archeozoic*, 79
Archetypes, 104
Artifact-making, 11–13, 15, 19, 27, 28, 188


© Springer International Publishing Switzerland 2015

M. Barbieri, *Code Biology*, DOI 10.1007/978-3-319-14535-8

Artifacts, 11, 12, 19, 28, 124, 184, 185, 188
Artificial messenger(s), 37
Association areas, 116
Atmosphere, 77
Automatic pilot, 114, 125
Automaton, 70, 173, 174
Autopoiesis, 90, 171, 175, 176, 187
Autopoietic system(s), 90, 175, 176
Axis genes, 102
Axolot, 134
Axons, 87, 115, 119

**B**
Bacteria, xiv, xvi, 6, 36, 49, 65, 72, 78–82, 84, 87, 89–91, 162–163, 181
Banded Iron Formations, 77
Basophiles, 82
Behaviour, xii, 66, 81, 84, 85, 114, 125, 131, 138, 145–146, 158, 162–163, 176–177
Bilateral symmetry, 106
Bilaterial, 106
Binomial terminology, 104
Biochemistry, 9, 11, 14, 31, 41, 44, 53
Bioelectric code, xii
Biohermeneuthics, 159, 162
Biolinguistics, 131–133
Biological code view of language, 133, 146
Biological complexity, 40, 183
Biological information, xv, xvi, 3, 5, 6, 9–11, 19, 21, 34, 179
Biological machine, 172
Biological meaning, xv, xvi, 34, 96–97, 162, 179
Biological sequence, 15, 22, 25, 28
Biological specificity, 10, 11, 20, 21, 36, 43, 57, 61, 63, 72
Biological symbolism, xi
Biosemiotics, viii, xvi, 131–133, 151–169
Biosphere, 90, 131, 159
Birth canal, 134, 136
Bithorax, 101
Blastocyst, 99
Blastomers, 99, 100
Blastopore, 98
Bluebacteria, 77
Body axes, 102, 106–109
Body plan, xi, 95–98, 101–102, 104–110, 181, 185, 188
Body segments, xi, 107
Bondmakers, 13, 14, 58, 61, 180, 181, 188
Brain, 30, 31, 41, 44, 45, 93, 95, 99, 109–127, 131, 133–138, 140–146, 163, 165, 181
development, 118–121
macroevolutions, 126–127
wiring, 134, 135, 137, 140
Brain–mind relationship, 111, 121–122
Bromodomains, 47
Bytes, 7, 8

**C**
Cadherins, 120
Calcium ions ($Ca^{2+}$), 43, 44
Cambrian, 75, 102, 103, 105–107
Cambrian explosion, 78, 91, 105–107, 181
Camera-eye, 116
Cardia bifida, 129, 136–137
Catalysts, 11, 12, 36, 40, 45, 167
Catalyzed assembly, 40
Catalyzed processes, 11
Category, 15, 67, 73, 96, 104, 113, 125, 138, 139, 153, 155, 164–166, 183
Causality, 138, 139
Cell death, 119, 120
Cell division, 79, 95, 100, 118–119
Cell fate, 98, 99, 121
Cell memory, 46, 98–99, 106
Cell pattern, 102, 106
Cellular level, 37, 75, 76, 153
Cellulose, 36, 49, 50
Cell wall, 50, 52
Central dogma, 36, 37
Centrioles, 48
Centromeres, 86
Centrosome, 49, 86
Cerebra bifida, 129, 136–137
Cerebral cortex, 141, 142
Channel code, 6, 7, 52
Chaos theory, 5, 16
Chemical energy, 17
Chemical evolution, 9, 11, 12, 57–59, 64
Chemical machines, 4
Chemical paradigm, 3–6, 8–11, 13, 15, 16, 19, 27
Chemistry, viii, ix, xii, xv, 3–17, 19, 24, 28, 33, 34, 53, 64, 84, 95, 132, 153, 155, 156, 188
Chemoaffinity hypothesis, 120
Chicken-and-egg paradox, 68
Chimpanzees, 136
Chloroplasts, 78, 79, 90
Cholecystokinin, 44
Chordata, 98
Chorion, 100
Chromatin, 45, 46, 53
Chromatin folding code, 52

Chromosome(s), 20, 45, 60, 78, 82, 84–86, 99, 101
Chromosome theory of heredity, 84
Cilia, 48, 49
Clades, 104
Cladistics, 29, 54, 104
Classical physics, 16, 27, 28, 159, 187
Classification, 52–53, 104–105
Classification of signs, 155
Code ambiguity, 69, 70
Code biology, xvi, 54, 152, 162, 169, 171–189
Code biosemiotics, 162
Code conservation, 72, 175
Code exploring, xv, 175
Codemakers, 12–14, 26, 30–32, 74, 172, 188
Codemaking, 31, 32
Code optimization, 38, 39, 71
Code paradigm, xv, 19–34, 54
Codepoiesis, xv, 72, 90–91, 175–176, 180, 187
Code-script, 20, 21
Codes of language, 146–147, 152
Codes of storytelling, 147
Codes of the body plans, 108–110, 181, 188
Code theory of mind, 111, 124
Code view of language, 146
Codeword(s), 37
Codified assembly, 40
Coding, xi, 3, 19, 35, 59, 75, 107, 111, 132, 151, 171
  rules, xii, xiv, xv, 11, 12, 14, 15, 22, 24, 26, 28, 33–35, 38, 43, 51, 53, 71, 83–85, 117, 120, 121, 123, 124, 133, 142–143, 146, 175, 179, 183, 185, 188
  space, xiv, 89, 90
Codon(s), xiii, 26, 31, 36–39, 41, 47, 52, 65, 68, 69, 72, 83, 162, 167
Cognitive development, 139
Cognitive system, 125–127, 129, 130, 134–137, 144–145, 188–189
Collagen, 104
Colonial organisms, 106
Combinatorial codes, 41, 47, 107, 108, 117, 121
Combinatorial gene expression, 108
Common ancestor, xiv, 38, 66, 68–70, 72, 75, 80, 81, 90, 91, 103, 104, 106, 107, 175, 188
Communication, 6, 7, 49, 50, 112, 119, 129, 130, 139, 156–158, 185
Communication theory, 52
Compartment codes, xi, xiv, 87–88, 90
Compartments, 60, 62, 78, 86–88, 90, 137, 175, 184
Complexity, xiii, xiv, 4, 13, 14, 19, 21, 29, 33, 37, 40–42, 45, 51, 52, 59, 61, 63, 82, 87, 89–91, 93, 94, 98, 109, 110, 123, 132, 136, 142, 145, 153, 163, 164, 174, 181–184
Complexity theory, 5
Computable, 24, 28
Computational theory, 121
Computer, 5, 7, 8, 10, 16–17, 73, 121, 122, 172
Computer science, 173
Connectionist theory, 121–122
Conscious brain, 115, 122
Conscious instincts, 115, 126
Consciousness, 113, 115, 122–124, 129
Conservation, xv, 45, 72, 175, 176, 180, 182
Constraints, 21–23, 52, 108, 134, 136, 153, 154, 161, 179–180
Consumers, 85
Context, 45, 142
Context-dependent behaviour, 81, 158, 162–163
Controller, 173, 174
Conventional rules, 28–30, 109, 120
Conventional signs, 32, 33
Conventions, 30, 32, 35, 87, 88, 105, 109, 178, 179, 182
Convergence, 96–97, 110
Convergent increase in complexity, 93, 94, 110
Copernican system, 27
Copying, xv, 6, 12, 13, 15, 19, 22, 24–28, 33, 34, 58, 69, 89, 90, 172, 177–181, 187, 188
Copymakers, 12–14, 24–26, 58, 181, 188
Corpus callosum, 141
Correspondence, xii, xiii, 26, 30, 31, 36–38, 40, 43, 83, 88, 154, 162, 182, 183
Cortical areas, 115, 116, 141
Cortical cells, 115, 141
Creole languages, 140
Cultural codes, 28, 35
Cultural semiosis, 28, 165, 168
Cultural signs, 155, 165
Culture, xi, 11, 28, 30, 127, 129, 131, 151, 157, 165, 166, 172, 187
Cyanobacteria, xiv, 77, 81
Cyanozoic, 77–79
Cybernetic brain, 114, 115, 125, 126
Cyclic AMP, xiii, 42–44
Cytoplasm, 36, 49, 51, 78, 85–86, 88, 98
Cytoskeleton, 78, 85–87, 90, 184
Cytoskeleton codes, xi, xiv, xvi, 86–87, 90, 181, 183
Cytosol, 88

**D**
Darwinian biosemiotics, 154–155
Darwinian evolution, 3
Darwinian threshold, 80
Decoding, 50, 65, 125, 141, 163–167
Deduction, 125, 155, 164, 185
Deep homology, 103
Definition of language, 130–131
Definition of semiosis, 28, 155–158
Degenerate, 38
Deoxyribonucleic acid (DNA), 6, 36, 37, 46, 52, 53, 63, 64, 78, 79, 81, 82, 84, 85, 102, 184
Derived quantities, 25
Design, 37, 81, 93, 94, 96, 97, 107, 109, 116, 141, 168, 173, 185
Determination, 98–99, 107, 117, 121, 161
Determinism, 16, 29, 159, 184, 185
Development, xii, 4, 20, 40, 64, 79, 93, 112, 129, 151, 171
Diacylglycerol, xiii, 43
Diamond code, 39
Dichotomy, 79, 151
Dictyostelium discoideum, 42
Differentiation, 41, 46, 86, 95, 97, 99, 100, 102, 106, 117, 120, 123, 163, 165, 188
Digital, 7, 9, 12–13, 107, 167
Diploblasts, 106, 107
Discontinuity, 4, 105, 166, 182
Distal-less, 102, 103
Distributed language, 141
Diversification, 69, 70, 120
DNA. *See* Deoxyribonucleic acid (DNA)
DNA structure/shape code, 53
Domain(s), xiv, 41, 42, 47, 49, 51, 72, 80–82, 102, 121, 168, 181, 183
Dorsal cord, 97, 98, 108
Double helix, 14, 36, 45
Dreams, 123, 124
*Drosophila*, 98, 101, 102
Dynamic instability, 48, 86, 87

**E**
Earth, 1, 3, 9, 13, 14, 28–29, 51, 57, 58, 60, 62, 64, 67, 68, 70, 73, 75, 77, 78, 82, 89–91, 180, 185, 188
Ectoderm, 95, 97, 100, 118
*Ediacara fauna*, 105, 107
Editing codes, 83
Editosomes, 83
Effector proteins, 46–47
Efficient causes, 93
Electrical signals, 112–118, 123
Embranchement, 96
Embryonic development, 40, 46, 79, 93–98, 100–104, 106, 108–110, 117, 118, 120, 134, 136, 142, 175, 177, 179
Embryonic fields, 99–101
Embryos, 93–110, 181, 188
Emergence, 95, 122, 124
Emergence theory, 122
Emotions, 113, 122, 126, 141
Encephalins, 45
Endoderm, 95, 100, 106
Endoplasmic reticulum, 78, 88
Energy, 4, 5, 17, 27, 34, 41, 43–45, 49, 58, 64, 73, 86
Enzyme(s), xiii, 36, 37, 44, 50–51, 59, 66–67, 72, 85, 174
Eons, 75, 77, 79
Epigenesis, 93–95, 109
Epigenetic code, xi, 46, 107
Epigenetic modifications, 49, 50
Epigenetic processes, 136
Error catastrophe, 60, 61
Error-correcting-codes, 7
Eubacteria, xiv, 80, 82
Eucarya, xiv, 72, 80, 81
Eukaryotes, 45, 53, 71, 78–80, 82, 83, 87, 89, 90, 181
Eukaryotic cell(s), 45, 48, 80, 86–88, 91, 112, 181, 183, 188
EvoDevo, 103, 177
Evolutionary strategies, xiv, 81–82
Evolution by natural conventions, 178
Evolution by natural selection, 178
Evolution of the brain, 112, 113, 115, 163, 181
Exons, 40, 41, 82, 83
Exploratory strategies, 87
Expressed information, 15
Extended mind, 155, 156, 168
Extensions of the modern synthesis, 176–177
Externalism, 171
External language, 130
Extracellular matrix, 50
Extra-uterine development, 129, 134, 135
Extremophiles, 82
Eye, 75, 94, 97, 103, 105, 116, 120, 139, 142
Eyeless, 102, 103

**F**
Faculty of language, 130, 132, 133, 136, 144
Faculty of language in the broad sense (FLB), 130, 131
Faculty of language in the narrow sense (FLN), 130, 131

Feedback, 42, 125
Feeling brain, 114, 115, 125, 126
Feelings, 16, 111, 113–115, 117–119, 121–126, 139, 156, 160, 161, 163, 164, 188
Feral children, 140
Fetalization, 129, 134–135
Fetalization theory, 133, 134
Final causes, 93
First cells, xiv, xv, xvi, 3, 11–13, 57, 65, 72, 73, 76, 80–82, 85, 88, 90, 152, 175, 176, 181
First common ancestor, 68–70
First messengers, xiii, 43, 45, 117
First neural cells, 122
First-person experiences, 111, 122–123
Fluid genome, 84, 85
Foetal development, 100, 129, 134–137, 144, 188
Foetus, 101, 133–135, 188
Folding, 45, 52, 78, 95, 174, 185
Following response, 138, 140, 142
Fossilized microorganisms, 75
Fossil record, xiv, 78–79, 105, 181
Fossils, 75, 77, 79, 95, 105, 106
Free will, 113, 129
Frozen accident(s), 21, 153–154
Functional information, 15
Functionalism, 5–6, 16, 70, 119, 136, 137, 161, 171
Fundamental laws, 21, 153
Fundamental observable, xvi, 22, 25, 27, 28, 34
Fundamental quantities, 25, 34
Fungi, 79

G
Gastrula, 98
Gene-centred view, 177
Generation, xiv, xv, 6, 9, 20, 25, 27, 40, 42, 44, 46–47, 49, 61, 63, 64, 69–72, 79, 84, 89, 93, 98–100, 106, 108, 114, 115, 119, 120, 123, 124, 126, 130, 132, 135, 136, 143, 158, 173, 175, 176, 179, 184, 185, 188
Genes, xv, 5, 19, 37, 57, 79, 97, 112, 135, 153, 172
Genetic code, 3, 19, 35, 59, 75, 108, 111, 132, 151, 171
Genetic information, xv, 6, 14, 15, 34, 37, 64, 65, 85
Genetic programme, 3, 5, 94, 109
Genome, xi, xvi, 5, 6, 46, 50, 52, 53, 81, 84–86, 89, 103, 184
Genome segmentation code, 53
Genotype, 5, 73, 74, 153, 174, 175
Genotyping, 8
Geological age, 75
Geological period, 75
Germ layers, 95, 100, 106, 118
Gestation, 134
Glucose, 42, 44, 49, 51
Glycomic code, xi, 51–52
Glycoproteins, 49, 50
Glycosylation, 49
Gödel's theorem, 9, 16
Golgi apparatus, 87, 88
Gray code, 107
Growth factors, xiii, 43, 44, 119, 143

H
Halophiles, 82
Hardware-software, 5, 121
Hemicelluloses, 51
Hensen's node, 98
Hereditary information, 5, 7, 36
Hereditary substance, 20
Heredity, 6, 9, 19, 36, 58, 73, 74, 84, 177, 179
Hermeneutic biosemiotics, 159–160
Heterochrony, 134
Heterogeneous nucleosoids, 64
Heterosoids, 64–65
Hierarchical view, 177
Higher-level brain processes, 111
Highway code, xii
Hippocampus, 117
Histological memory, 99
Histone code, xi, xiv, xvi, 45–48, 110, 181
Histone modifications, 46–48
History of life, xi, xv, 14, 33, 45, 54, 57, 59, 67, 68, 72, 75, 86, 90, 113, 123, 125, 129, 144, 159, 160, 163, 180–182
History of the Earth, 75, 79, 105
Homeobox, 102, 103
Homeodomain transcription factors, 121
Homeotic genes, 101–102
Homologous genes, 103, 177
Homology, 103
Horizontal gene transfer, 63, 80, 89
Hormones, xiii, 42–45, 50, 163
Hox codes, 107–109
Hox genes, 46, 102–104, 107–109
Humanities, 134, 151, 160
Human semiosis, 29, 172, 186
Hydrogen bonds, 20
Hydrophilic, 66, 67

Hydrophobic, 66, 67
Hylozoism, 156
Hypercycle, 60

**I**
Icons, 32, 33, 143–146, 157, 186
Immanent design, 94, 97, 109
Immune system, 49, 120
Inanimate matter, 3–4, 13, 14, 20, 22
Independent worlds, xii–xiii, 26, 29–30, 35, 43, 47, 83, 108, 109, 162
Indexes, 32, 143–146, 157, 186
Induction, 97–99, 101, 125, 155, 164, 185–186
Information, xv, 3–17, 19–22, 24–27, 33, 34, 36, 37, 42–45, 48, 50, 52, 53, 63, 72, 73, 81, 110, 116, 117, 123, 124, 126, 139, 153, 154, 159, 160, 172, 174, 175, 179, 181, 188
  paradigm, xv, 3–6, 8–10, 15, 16, 19, 27, 34
  quantity, 6
  theory, 6–7, 9
Innateness, 132
Innenwelt, 126, 131
Inositol trisphosphate (IP3), xiii, 43
Instinctive brain, 114–115, 126
Instincts, 111, 113, 115, 117, 118, 121, 122, 124–126, 156, 188
Insulin, 45
Intelligent design, 168
Intentional information, 15
Intercellular communication, xi
Intermediate brain, 113–115, 117, 123–126, 146
Intermediate filaments, 78, 86
Intermediate neurons, 113, 114, 188
Internalism, 171
Internal language, 130
Internal spatial map, 117
Interpretability, 132, 159
Interpretant, 156–158
Interpretation, 32, 76, 115, 125, 126, 132, 133, 142, 143, 151, 152, 155, 157–160, 162–168, 180, 185–186
Interpretive brain, 125
Interpretive semiosis, 132, 151
Intracellular receptors, 43
Intrauterine development, 135, 137
Introns, 40, 41, 53, 82, 83
Invariants, xv, 139, 189
Invertebrates, 106, 126
Iron bands, 77

**J**
Juvenile ape, 133–134

**K**
Kin selection, 176

**L**
Labels, 23, 24, 96, 109
LAD. *See* Language Acquisition Device (LAD)
Language, xvi, 7, 8, 10, 23, 26, 29–31, 35, 46, 117–118, 126, 129–147, 152, 155, 157, 166, 178, 182, 183, 185–187
  codes, 182, 183, 187
  genes, 135–136
Language Acquisition Device (LAD), 132
Last common ancestor, 68–70, 72
Law of mass action, 9
Laws of forms, 104
Laws of physics, xii, 5, 20–22, 93, 154
Learning, xvi, 23, 32, 85, 120, 125, 131, 138, 140, 143–145, 157, 164, 167, 186
Lectins, 50
Linear information, 5, 73, 174
Linguistics, 8, 10, 15, 19, 130, 140
Lipids, 64, 83
Lipid world, 64
Living systems, 3–5, 11, 12, 17, 19, 20, 22–24, 27, 28, 34, 35, 38, 59, 73, 110, 133, 152, 156, 163, 164, 166, 167, 169, 175, 176, 179, 180, 183–186, 189
Logic, 5, 9, 16, 86, 87, 89, 109–110, 125, 142, 153, 155, 164, 174, 176, 184, 187, 188
Lower-level brain processes, 111, 121
Lysosomes, 78, 87

**M**
Machine, 4–6, 10–17, 20, 21, 24, 28, 33, 37, 40, 41, 57–59, 65, 73, 83, 88, 110, 130, 160, 172–176, 180, 181, 188
Macroevolution, xiv, xvi, 81, 85, 91, 109, 123–127, 129, 147, 160, 164, 177, 180–182, 187, 189
Major transitions, 33, 72, 83, 86, 108, 125, 127, 177, 180–182
Mammals, 96, 98, 99, 102, 116, 134–136, 164
Manufactured molecules, 11, 13, 14, 57, 124, 188
Manufactured objects, 11, 13

Manufacturing code(s), 111, 124, 146, 183, 184
Mapping, xii, xiii, xiv, 16, 21, 35, 47, 52, 53, 68, 84, 108, 109, 117, 123, 146, 153, 172
Mathematical models, xii, 16, 160, 187
Mathematical principles, 5, 171
Meaning, xii, 3, 19, 35, 60, 75, 96, 114, 130, 151, 177
Meaningful information, 15
Mechanical energy, 17
Mechanical stimuli, 118
Mechanism, 4, 5, 10, 15–17, 39, 41, 46, 49, 58, 59, 61, 63, 72, 79, 83, 85–88, 93, 94, 97–99, 101, 115, 117–120, 122, 124, 125, 131, 132, 143, 155, 160, 161, 171, 175–181, 185–187
Mechanisms of evolution, 177–178
Mechano-receptors, 118
Meiosis, xvi, 79, 86
Membrane potential, 112, 113
Membranes, 43, 49, 64, 78, 81, 83, 85, 87, 88, 90, 93, 112, 113, 117, 120, 184
Memory, 8, 32, 46, 85, 98–99, 106, 110, 119, 125, 139, 164
Mendelian behaviour of genes, 84, 85
Mental categories, 138, 139
Mental images, 30, 115–116, 121, 138, 143, 157, 163
Mental objects, 26, 30–31, 138, 162
Mesoderm, 95, 97, 100, 106, 118–119
Messenger(s), xiii, xiv, 37, 40, 43–45, 47, 81, 82, 117, 183
Messenger-RNAs, 31, 37, 40, 61, 65, 66, 69, 70, 72, 82, 83, 90, 183
Metabolic code, xi, 41–42
Metabolism, 41, 42, 45, 58, 59, 72–74, 85
Metaphor(s), 10, 11, 13–14, 19, 20, 33, 152
Meteorites, 57, 64, 67
Methylation, 45–48
Micro-compartments, 62
Microenvironments, 67
Microevolution, 177
Microfibrils, 51
Microfilaments, 78, 86
Microfossils, 76, 78
Microorganisms, 39, 42, 51, 75–78, 158
Micro RNAs, 83
Microscope, 75, 76, 94, 97, 101
Microtubules, 48, 49, 78, 86, 87
Midbrain, 115, 116, 141
Mind, xiv, xvi, 8, 23, 26, 31, 32, 41, 52, 76, 86, 106, 110–127, 130, 133, 138, 139, 141, 145–147, 152, 155, 156, 162, 165, 168, 179, 181, 188
Mitochondria, 78, 79, 88, 90
Mitosis, xvi, 46, 79, 85, 86, 99
Mitotic spindle, 48, 49, 85, 99
Model(s), xii, 4, 5, 16–17, 27, 29–30, 35, 36, 41, 53, 60, 73, 107, 110, 122, 123, 125, 126, 129, 132, 135–137, 145, 146, 155, 160, 161, 165, 167–169, 173–176, 187
Modelling system, 126, 131
Modern biology, 5, 15, 26, 73, 153, 175, 187
Modern genetic code, 68–73
Modern synthesis, 3, 6, 154, 155, 159, 171, 176–177, 180, 187
Modulation code, 53
Molecular
  adaptors, 12, 40, 47, 83, 86
  artifacts, 11–13, 28
  biology, xiv, 3, 9–11, 14, 20, 28, 36, 49, 73, 75, 151, 153, 171, 187
  biosemiotics, 151, 153, 158
  coding, 27, 28, 33
  copying, 6, 13, 27, 33, 58, 177, 178, 187
  determinants, 108, 109
  embryology, 102–103
  machines, 11–14, 20, 28, 33, 37, 57–59, 73, 83, 180, 188
  phylogeny, 60
  semiosis, 28, 33, 151
  worlds, 47
Monoblasts, 106
Monozygotic twins, 100
Morphogenesis, 48, 95, 105, 121
Morse code, xii, 20, 26, 39–40, 162, 172
Morula, 99, 100
Mosaic determination, 98
Motor neurons, 113, 114, 119
Motor organs, 112–115, 123
Multicellular life, xvi, 79, 91
Multicellular systems, 123, 125, 164, 183
Multifunctional molecules, 44, 45
Multi-gated connections, 114
Mutation(s), 72, 84, 101, 103, 136, 176, 179–180, 182

**N**

Names, 8, 14, 23–24, 29, 32, 36, 37, 48, 50, 52, 62, 73, 76, 88, 89, 98, 105, 126, 131, 133, 140, 143, 147, 151–154, 157, 158, 162, 165
Natural conventions, 33, 87, 88, 178, 179, 182

Natural selection, 3, 6, 16, 19, 54, 58, 103, 138, 143, 154–155, 160, 172, 176–180, 182, 184
Natural signs, 32, 33, 155, 165
Necessity, xiv, 8, 15, 21, 23, 30, 34, 35, 38, 41, 47, 48, 66, 69, 72, 93, 97, 109, 123, 125, 133, 145, 153, 154, 156–158, 161, 165, 173, 176–178, 187
Neocortex, 116
Neoteny, 133, 134
Nerve growth factors, 119, 143
Nervous system, 45, 112, 113, 117–122, 132, 142, 143, 163, 185
Neural cells, 46–47, 112, 115, 121, 164
Neural code, 31, 111, 117–118, 124, 163, 181, 182, 187, 188
Neural code for taste, 117
Neural computations, 122, 124
Neural development, 119, 142–143
Neural networks, 114, 122, 124–126, 164
Neural representation(s), 116, 164
Neural signalling, 116, 122, 123
Neuroblasts, 118–119
Neurochemicals, 119
Neuron firings, 31, 111, 181
Neurons, 87, 112–114, 117, 119–121
Neurotransmitters, xiii, 44–45
New laws of physics, 20–22, 153, 154
Newton's greatest discovery, 21, 154
Nominable entities, xv, 23–26
Nominal entities, 23, 24
Non-equilibrium thermodynamics, 5, 16
Non-Mendelian heredity, 84
Nuclear signalling code, xi
Nucleic acids, 10, 12, 13, 36, 41, 58, 89, 181
Nucleoli, 62
Nucleosoids, 61–64
Nucleosome(s), 45, 46, 52, 53
Nucleosome positioning code, 52
Nucleotides, xiii, 5–8, 26, 31–33, 36–39, 42, 50, 58, 59, 61, 65, 66, 83, 102, 142, 162, 188
Nucleus, xvi, 36, 78, 83, 88

**O**

Objective sequence, 24, 31
Observables, xv, 22–25, 27, 28, 34
Observer, 23, 26, 123
Odorant receptor code, 117
Ontogenesis, 118, 132
Ontogenetic learning, 138
Ontogeny of mind, 139
Ontological claim, 9–11
Ontological entity, 19
Operative definition, xii–xiii, 25
Optic chiasm, 115, 116, 141
Optic cup, 97
Optic nerve, 97, 115, 116, 120, 141
Optic vesicle, 97
Optimization, 38, 40, 71
Organelles, 48, 49, 78, 112, 175
Organic codes, xi, xii, xiv, 20, 26, 34–54, 72, 75, 81, 86, 89–91, 108–110, 120, 121, 132, 160, 163, 165, 166, 172, 176, 178–185, 187, 189
Organic hardware, 73
Organic information, 15, 19, 22–26, 28, 34
Organic meaning, 15, 19, 22, 23, 26–28, 32, 34, 47, 48, 160, 162
Organic semiosis, 29, 33, 132, 133, 162–164, 166, 186
Organic software, 73
Organizer, 98
Organogenesis, 101
Origin of animals, xvi, 106–108, 125, 163, 186
Origin of interpretation, 182
Origin of language, xvi, 129, 133, 136, 137, 147, 155, 182, 187
Origin of life, xi, 3, 9–11, 13, 33, 57, 60, 76, 81, 89, 109, 111, 132, 133, 147, 152, 163, 187
Origin of man, 133, 134, 145
Origin of mind, xvi, 111, 133, 147, 181
Origin of the nucleus, xvi, 83
Osmotic pressure, 112
Oxygen revolution, 77, 78

**P**

Paedogenesis, 133
Paedomorphosis, 133
Paleontology, 75, 95, 182
Pansemiosis, 156
Paradigm(s), xv, xvi, 4, 5, 8, 9, 15, 19, 27, 34, 158, 160, 166, 168, 171
Pectins, 51
Peircean biosemiotics, 152, 158–159, 167, 168
Peircean semiosis, 159
Peirce model of semiosis, 132
Peptide(s), 44, 60, 61, 67, 88
Peptide bonds, 24, 61, 65, 66
Peptidyl transferase center, 65
Perceptions, 113, 116–117, 126, 138, 164
Periodic table, 132, 133
Phanerozoic, 79
Phase-transitions, 5, 122
Phenotype, 5, 73, 74, 121, 153, 174, 175

Phenotype-genotype duality, 5, 73
Philosophy, 23, 53–54, 152, 155, 159
Phonetics, 130
Phosphodiester bonds, 61
Phosphorylation, 46, 48
Photoreceptor cells, 115, 116
Photosynthesis, 64, 77
*Phycoides pedum*, 105
Phyla, 96, 103, 177
Phylogenetic learning, 138
Phylogenetic trees, 29, 70
Phylotypic stage, 96, 97, 104, 108
Physical biosemiotics, 153–155
Physical information, 15
Physicalism, 10, 11, 16
Physicalist thesis, 10–12, 14, 15, 33
Physical laws, 21, 132, 142, 153, 154, 171, 179
Physical necessity, xiv, 40, 48, 50, 109, 143, 178
Physical quantities, 4, 10–12, 14–16, 19, 24, 25, 33, 34, 73
Physical theory, 20–23, 153, 154
Physics, xii, 4, 5, 16, 20–25, 27, 28, 34, 93, 132, 133, 153, 154, 159, 179, 180, 187
Physiosemiosis, 156
Phytosemiosis, 158
Placenta, 99, 100, 102
Plants, 49, 51, 52, 78, 79, 84, 89, 104, 146, 158
Plasma membrane, 78, 85, 87
Platonic world, 23
Pleated sheet, 66, 67
Pole plasm, 98
Polymerases, 33, 58, 174
Polymerizing ribosoids, 61
Polymers, 5, 13, 50, 51, 58, 180–181
Polynucleotides, 58, 61, 181
Polypeptides, 13, 37, 58, 61, 65, 67, 174, 180–181, 185
Polysaccharides, 13, 51, 58, 181
Population genetics, 54, 103, 171, 176, 177
Positional determination, 98
Postchemical evolution, 57
Post-translational modifications, 45–46
Pragmatism, 155
Prebiotic chemistry, 64
Precambrian, 75–78, 85, 86, 90–91, 105
Precambrian evolution, 75, 79
Precambrian life, 75–77
Precambrian microorganisms, 75, 76
Precellular evolution, 60
Precellular systems, 66
Preformation, 94
Prezoic, 79
Primary amino acids, 14, 67
Primary kingdoms, xiv, xvi, 70, 72, 79–81, 181
Primary transcripts, 40, 82, 83, 90, 183
Primates, 96, 133–135
Primitive Earth, 3, 13, 14, 57, 58, 60, 62, 64, 67, 70, 73, 180
Primordial cells, xiv, 41, 88
Principle of continuity, 156
Principle of Least Action, 132, 133
Principle of the correlation of parts, 96
Probability function, 52
Producers, 85
Prokarya, 89
Prokaryotes, 53, 71, 78–80, 183
Propositional thinking, 139
Proteins, 5, 19, 36, 57, 82, 98, 111, 146, 165, 172
Protein synthesis, xiii, 5, 12, 30, 31, 33–37, 40, 41, 62, 65–73, 80–83, 98, 146, 153, 165, 178, 181, 183
Proterozoic, 78, 79
Protista, 78–79, 89, 90
Protocadherins, 120
Protoribosomes, 61
Pseudo genes, 41
Psychology, 113, 156
Psycrophiles, 82
Ptolemaic model, 27

**Q**

Qualities, xv, 22, 23, 25
Quantities, xv, 4, 6, 8, 10–12, 14–16, 19, 22–25, 27, 33, 34, 64, 73, 188
Quantum effects, 122
Quantum numbers, 24
Quantum theory, 16, 153
Quasi-replication, 61–64

**R**

Radiata, 106
Random drift, 176
Random polymers, 58
Rational faculties, 145–146
Reality, xi, 16, 21, 24, 26–28, 39, 47, 52, 94, 123, 126, 132, 138, 143, 151, 153, 157, 163, 172, 182, 187
Receptor(s), xiii, 43, 117, 167, 183
Recognition processes, xiii, 39, 40, 43, 48, 83, 87
Recursion, 35, 130
Reductionism, 6, 16, 70, 80, 95
Redundant, 38, 73
Reflex arch, 113, 114

Regulation, xvi, 41, 42, 45–47, 53, 81, 83, 174, 184
Regulatory codes, 52, 184
Replication paradigm, 59–60
Repression, 46, 47
Respiration, 64
Retina, 97, 115, 116, 123, 126, 141, 142
Ribogenes, 73
Ribogenotype, 73
Ribonucleic acid (RNA), 31, 36, 37, 40, 59, 61–65, 68, 69, 71, 72, 80, 82–84
  codes, 82–84
  editing, 83
  polymerases, 40
  synthetases, 71
  world, 59, 64
Ribonucleoprotein granules, 36, 37
Ribonucleoproteins, 30, 61, 62, 74, 175
Ribopeptides, 60
Ribophenotype, 73
Ribosoids, 60–64, 73
Ribosomal proteins (r-proteins), 65, 69–71
Ribosomal-RNAs, 37, 60, 61, 65, 66, 68, 80
Ribosome(s), 37, 40, 59, 61, 65, 68–71, 79, 83, 88, 167, 174
Ribotype, 60, 73–74
Ribozymes, 59, 73

S
Scanning factors, 37
Schools of biosemiotics, 151
Schrödinger's prophecy, 20–21
Scientific modelling, 16, 161
Secondary amino acids, 14, 67
Second messengers, xiii, 43–45, 47, 117, 183
Segment genes, 102
Self-assembly, 61, 62, 64, 69, 70, 174–175
Self-fabrication, 171, 175
Self-regulation, 99–101
Self-replicating machines, 21
Self-replicating molecules, 59, 60
Self-replicating system, 21, 60, 153
Self-replication, 22, 59
Self-reproducing automata, 152
Self-reproduction, 64, 153, 154, 173
Semantic biology, 162
Semantic information, 15
Semantics, 8, 130
Semiosis, 28–33, 125, 132, 133, 143, 151–159, 162–166, 168, 169, 172, 184–186
Semiosphere, 131, 159
Semiotics, xvi, 29, 126, 151, 155, 158–160, 166–168
Semiotic system, 29, 30, 32, 152, 172
Sensations, 113, 116–118, 122, 126, 163, 181
Sense organs, 112–118, 123, 124, 126, 163
Sensory inputs, 114, 163, 164
Sensory neurons, 113, 114
Sequence codes, 52–53
Sequence information, 8, 34
Sequences, xi, xiii, xvi, 5–15, 22–26, 28, 31–34, 36, 37, 41, 50–53, 60, 65–69, 71, 72, 82–84, 88, 93, 95, 97, 102, 108, 111, 114, 124, 132, 135, 139, 159, 174, 179, 184, 185, 188
Servo-mechanism, 115
Shark Bay, 76, 77
Signa data, 32
Signal integration, 45
Signal integration codes, 44–45
Signalling, 44, 81, 112, 120, 122, 181, 183, 184
Signalling codes, xi, 183, 184
Signal transduction, xiii, xiv, 43, 44, 64, 80, 112, 117, 123, 163, 167, 181, 183
Signal transduction codes, xi, xiii, 43–45, 47, 72, 81, 117, 163, 181, 182, 188
Signa naturalia, 32
Signa symbolica ex cultura, 165
Signa symbolica ex natura, 166
Signs, xii, 21, 28–33, 45, 47, 48, 76, 77, 109, 143, 145, 151, 152, 154–159, 163–167, 186
Small interfering RNAs, 83
Small nuclear RNAs, 83
Small nucleolar RNAs, 83
Small shelly fossils, 105, 106
Sodium channels, 113
Solenoids, 45
Sonic hedgehog (Shh), 109
Source code, 6, 7, 52
Space code, xiv, 53, 89, 90, 117, 188
Special constraints, 21–22
Species, 6, 29, 40, 45, 52, 61, 62, 71, 76, 79, 89, 95, 96, 102, 104, 106, 109, 117, 126, 127, 129–131, 133–135, 137, 138, 142, 144, 155, 158, 166, 177, 182, 183, 186, 188
Specificity, xii, 7–8, 10, 11, 13, 20, 21, 23, 36, 43, 51, 57, 61, 63, 72, 83
Speech, 130, 141, 145
Speech centres, 141
Spiegelman monster, 59

Spliceosomes, 40, 83
Splicing codes, xi, xiv, xvi, 40–41, 53, 83, 84, 90, 175, 181–183
Split-brain operation, 141
Spontaneous molecules, 11, 13
Spontaneous reactions, 9, 11, 12, 15, 57
Statistical information, 15
Statistical proteins, 65–69, 71, 72
Steam-engine, 4, 16, 17
Stereochemical hypothesis, 35, 39, 53
Stereochemistry, 39–40
Stony carpets, 76
Storytellers, 189
Storytelling, 147
Streamlining, xiv, 82, 89
Stromatolites, xiv, 76
Structuralism, 171
Subjectivity, 22, 23, 111, 117, 126, 131
Subunit(s), xiii, 5, 11, 22, 24, 25, 57, 65
Sugar-binding proteins, 50
Sugar code, xi, 49–50
Sugars, 45, 49, 50
Supramolecular systems, 37, 61, 66, 90, 147, 188
Symbiosis, 79, 90
Symbols, 7, 21, 32, 33, 42, 134, 139, 143–146, 152–155, 157, 165, 166, 173, 182, 186
Symptoms, 32, 165
Synapomorphies, 104, 105
Synapses, 112, 119, 120, 140, 143
Synaptic connections, 111, 119, 135, 143, 144
Syntactic information, 7, 8
Syntax, 130, 132, 133
Synthetases, xiii, 36, 37, 39, 68, 70, 71
Systema Naturae, 104
Systems Biology, 5, 171, 187

**T**

Tactile sensations, 118
Taxa, 48, 104–106, 177
Taxonomy, 104
Teleology, 10, 33, 34
Template, 11–13, 24, 25, 49, 57–59, 173, 181, 188
Template-dependent molecules, 11
Termination signals, 38
Territory, 16, 21, 123, 153
Theoretical framework(s), xv, xvi, 5, 19, 21, 34, 134, 154, 156, 171, 177, 179
Theory of mind, 111, 124, 145
Thermodynamic machines, 4
Thermodynamics, 4, 5, 9, 16, 27, 28, 188
Thermophiles, 82
Three-dimensional structures, 73, 86, 108, 110, 174, 185
Time-objects, 111
Tinman, 102, 103
Tommotian, 105
Trace fossils, 105
Transcription, 37, 40, 46, 53, 82, 83, 90, 121, 174, 183, 184
  codes, 53
  factor binding regulatory codes, 52
Transfer-RNAs (tRNAs), xiii, 37–41, 53, 65, 66, 68, 71, 72, 83, 142
Translation, 14, 31, 37, 38, 40, 41, 49, 53, 61, 69, 71, 73, 82, 83, 90
  errors, 38, 71
  framing code, 53
  pausing code, 53
Transmembrane system, 43
Transposons, 84
Tree of life, 89
Triadic process, 156
Trilobites, 105, 106
Triploblasts, 106, 107, 118, 126
Tubulin, 48, 49
Tubulin code, xi, 48–49
Turbo codes, 7
Twins, 27, 58–59, 100

**U**

Ubiquitin code, xi
Umwelt, 126, 131
Uncertainty, 3, 21, 103, 132, 154, 159
Unconscious brain, 115, 122
Unconscious instincts, 115, 126
Universal constructor, 173, 174, 176
Universal copier, 173
Universal grammar, 131, 133
Universal principles, 132, 133
Universe, 5, 21, 91, 126, 153, 156, 168, 178, 185, 187
Urkaryotes, 80, 82

**V**

Vertebrates, 96–98, 103, 104, 106, 118, 126, 136
Vesicles, 64, 87, 88, 93, 97, 112, 113
Viroidea, 89
Viruses, 6, 59, 63, 89
Visual cortex, 115, 116, 141
Visual information, 116, 123
Vitalism, 4, 15, 161
Von Neumann machine, 173–176

**W**
Web of life, 88–90
Wiring, 119, 120, 134–135, 137, 140, 143
Wobble hypothesis, 38
Wolf children, 140

**X**
Xyloglucan, 51

**Z**
Zoosemiotics, 151, 156–157
Zootype, 105

The manufacturer's authorised representative in the EU is Springer Nature Customer Service Centre GmbH, Europaplatz 3, 69115 Heidelberg, Germany. If you have any concerns regarding our products, please contact ProductSafety@springernature.com

Printed and bound by CPI Group (UK) Ltd, Croydon, CR0 4YY
30/09/2026
02225237-0004